THE NATURE OF
BIOCHEMISTRY

THE NATURE OF
BIOCHEMISTRY,

BY

ERNEST BALDWIN
B.A., PH.D., F.I.BIOL.

*Professor of Biochemistry at University College in the University of
London, formerly Fellow of St John's College, Cambridge
Fellow of the New York Academy of Sciences*

SECOND EDITION

CAMBRIDGE
AT THE UNIVERSITY PRESS
1967

Published by the Syndics of the Cambridge University Press
Bentley House, 200 Euston Road, London, N.W. 1
American Branch: 32 East 57th Street, New York, N.Y. 10022

© Cambridge University Press 1962

Library of Congress Catalogue Card Number: 66-23107

First published	1962
Reprinted	1965
Second edition	1967

Translations
  Danish   1964
  Japanese   1964
  Spanish   1965
  Dutch and
  Czechoslovakian
  (In preparation)

Printed in Great Britain
at the University Printing House, Cambridge
(Brooke Crutchley, University Printer)

Page 68

The formula for *ribulose*-1:5-*diphosphate* should read:

$$
\begin{array}{c}
CH_2O\ \textcircled{P} \\
| \\
CO \\
| \\
HCOH \\
| \\
HCOH \\
| \\
CH_2O\ \textcircled{P}
\end{array}
$$

Baldwin: Nature of Biochemistry, Second Edition

CONTENTS

v

To
PAULINE, NICOLA
AND NIGEL

PREFACE TO THE FIRST EDITION

If it is true that a good wine needs no bush, it may be similarly true that a good book needs no preface. I hope that this may be a good book, for it has been written for a specific purpose. It is meant to be read, not studied.

Degree courses in biochemistry are now offered in a good many colleges and universities. Year by year many school leavers apply for admission to these courses and yet, in my own experience, few of them start with more than a vague idea of what biochemistry is all about. It is for their enlightenment that this book has been made. Possibly there will be others who would find it a useful starting-point for further study.

I have tried to tell a story that is factually accurate with a minimum of oversimplification. If some passages are a little difficult to follow —and biochemistry *is* an intricate subject—I hope the fascination of the story will lighten the reader's task.

It has been difficult to decide what to include and what to omit because, in a book of this size, it is possible to skim over only a part of the whole story. It has been my ambition to keep the book small, for size is apt to be inhibitory to the hard-pressed student of today. In these days of over-early specialization in the schools I have no wish to add to the burden; that is why the book has been written for reading rather than study. It is not intended in any sense to be a supplementary or complementary treatise.

I should like to thank the many friends and colleagues who have given me valuable comment, suggestions and criticism, and special thanks go to Mrs Joan Rosemeyer for preparing the diagrams, and to Mrs P. H. Clarke, whose advice and opinion have been unstintingly given.

E. B.

LONDON
July 1961

vii

PREFACE TO THE SECOND EDITION

The preparation of this second edition has provided me with an opportunity to correct a few factual errors which escaped my attention in the original. I have at the same time modified a few passages which might otherwise have been interpreted as teleological. The whole text has been carefully scrutinized and revised and, in response to the suggestions of a few reviewers, a new chapter has been added so as to include a brief account of some of the major features of photosynthesis.

E. B.

LONDON
January 1966

ACKNOWLEDGEMENTS

The author wishes to thank Dr F. H. C. Crick, F.R.S., and Penguin Books Ltd for permission to reproduce Fig. 17 and Professor F. S. Fruton and John Wiley and Sons Inc. for permission to reproduce Figs. 7 and 9.

PROLOGUE

The scientific study of living things began with the straightforward observation and description of whole organisms and their visible parts. The only tools needed were the human eye and a few simple instruments—a pair of scissors and a sharp knife, perhaps. The systematic classification of plants and animals was certainly attempted by Aristotle and perhaps even before his time. This classification was based upon observable resemblances and differences between wholes and parts. Only one general property was at that time known to run right through; that all plants and animals are *alive*; they grow and reproduce.

Later on the invention and development of the optical microscope brought to light a further general feature that is common to all living things, namely that they are all built up from great numbers of microscopic units known as *cells*. The discovery soon afterwards of an immense realm of previously unknown, single-celled organisms—bacteria, protozoa, yeasts and other unicellular fungi—added to the older sciences of botany and zoology the newer science of microbiology.

Many inquiring minds remained unsatisfied with a knowledge only of structure as revealed by the eye, the knife and the microscope, because this provided an answer to only one of two fundamental questions that we all ask, practically from the cradle; we want to know 'What is it made of?' and 'How does it work?'. The problem of how living things work began to be studied fairly early in the history of biology because, once one knew how living things are built and of what parts they are made, it began to be possible to understand some at least of the ways in which they do the things they do. As a result biology began to separate into two main branches, the structural, or *morphological*, and the functional, or *physiological*.

In more recent times, and mainly in the last sixty years, new methods for the study of living things have been added to visual observation, dissection and microscopy, namely methods borrowed or adapted from physics and chemistry. Two new biological sciences have been born and are still actively developing, *biochemistry* and *biophysics*. This book is concerned chiefly with biochemistry, a science which aims at answering in chemical terms the questions 'What is it made of?' and 'How does it work?'. Whereas the eye works at the gross level of visible objects, the microscope reaches down to the finer levels of groups of cells, single cells, cellular inclusions as minute as cell nuclei and even smaller particulate materials. Biochemistry, however, works at a finer level still. It operates at a level that cannot be reached even by the most modern optical or even the newer ultraviolet and phase-contrast microscopes.

Biochemistry works, in fact, at the molecular level. Only the electron microscope can reach far down towards this level. It can tell us a good deal about the structure of natural molecules whose molecular weights are of the order of millions, but has the disadvantage that it cannot look at them if any water is present. And water is the most universally distributed of all biological substances. X-radiography again can pick out spatial relationships within large molecules provided they are more or less crystalline. But all these are essentially static observations. Only the application of chemical and physical methods can tell us very much about the dynamic and functional relationships existing between molecules whose molecular weights amount to a few tens, hundreds or thousands. And, water apart, it is of molecules of this order of size that living materials are for the most part composed.

Living organisms of one kind or other can be found in the sea, on the sea shore, in estuaries, swamps and marshes, in rivers, ponds, lakes, on dry land, and even in deserts and hot springs. Even the concentrated salt lakes of California harbour living inhabitants, and the insects have even contrived an inhabitant for petroleum pools. Yet one of the most striking discoveries of biochemistry has been that, in spite of the immense variety of size, shape, form and structure of living things as a whole, they all have a very great deal in common; far more in fact that any one would have dared to imagine, even after the cell theory had been propounded and had found general acceptance.

In this book we shall attempt to give an account in chemical terms of the materials from which living stuff is made and to give also an account of some of the chemical operations that underlie the manifestations and the maintenance of life itself.

Prologue

Students interested in biology and in chemistry will find in bio-chemistry a link between these two sciences, and this may help to vitalise and integrate what they have already learned. It may be, perhaps, that this first glimpse of biochemistry will frighten some, but perhaps it will encourage others to go further and learn more about the fascinating fields that biochemists have already charted and the newer territory that they are only now beginning to explore. To go into new and uncharted territory is always exciting, and in biochemistry we have to explore the innermost molecular territory of life itself.

THE CONSTANCY OF
THE INTERNAL ENVIRONMENT

INTRODUCTION

Biochemistry originated as an offshoot from human physiology when it came to be realized that the chemical analysis of urine, blood and other natural fluids can help in the diagnosis of this disease or that. In its early days biochemistry was accordingly known as Chemical Physiology. Now physiology, according to the *Concise Oxford Dictionary* is 'the science of normal functions and phenomena of living things'. Biochemistry is concerned particularly with the chemical aspects of these functions and phenomena and is but one of the many ways in which physiology can be studied. You can learn a lot about physiology by watching the monkeys at the zoo. Or you might take a cat or a dog to bits and study the behaviour of the bits and pieces. And you can go further still by extracting from those bits and pieces a variety of chemical substances, and by studying the chemical composition of those substances and the chemical interactions that take place between them.

If you watch the monkeys in the Zoo you are studying the physiology of behaviour. But if you study the behaviour of animal *molecules* rather than whole animals you could legitimately be called a chemical physiologist, or, in today's language, a biochemist.

THE ORGANISM AND ITS ENVIRONMENT

There are some people who still believe that the earth was originally populated by an act of special creation, strictly according to the book of Genesis, but the majority opinion today is that life began many millions of years ago and that from very simple and lowly beginnings the micro-organisms, plants and animals of the present day arose, slowly and step by step, by the painful process of evolution.

It is even possible to argue that life never had a 'beginning'; that life did not suddenly appear in a non-living universe, but was only the result of an increase in the degree of complexity and organization of something that already existed but was not 'living' in the sense that we understand 'life' and 'living' today. One thing is certain however; living things as we know them have changed a lot in the past, are still changing at the present time, and are likely to go on changing in the future.

The great French physiologist, Claude Bernard, was the first to point out that every animal lives in two distinct and separate environments, and not merely in one. There is first of all the outside world, the *external environment*, the properties of which can and do vary widely with the wind and the weather. Secondly, there is the animal's own personal, *internal environment* that it carries around inside it. This internal environment is represented by the animal's blood plasma and tissue fluids. Claude Bernard emphasized that *no matter how the external environment may change, the properties of the internal environment must be kept constant within very narrow limits*; otherwise the animal dies. This is well known; indeed, it is a matter of everyday laboratory experience. For example, if the blood of an animal becomes and is allowed to remain too acid or too alkaline, too dilute or too concentrated, too hot or too cold, death is the result. Most animals possess mechanisms for maintaining this vitally important internal constancy, no matter what happens in the outside world. We, in common with most mammals, maintain a virtually constant body temperature. If it falls we can generate heat and warm ourselves up again by shivering; and if it becomes too high, then we can lose heat by sweating. There are many other examples—for instance the excretion of an unusually alkaline urine after alkaline drinks have been taken.

THE EXTERNAL ENVIRONMENT

The earliest and most primitive of living things can hardly have possessed any of the many regulating devices that are found in modern animals. It follows that if internal constancy was as important then as it certainly is nowadays, these early living beings must have lived in an external environment of which the properties were relatively constant. Such an external medium would enable them to rely upon the constancy of the world outside for the internal constancy upon which their own continued existence depended. If such an external environment exists in Nature that, surely, must be the ancestral home of the living things that eventually evolved to populate the world of today.

Terrestrial environments of the kind with which we ourselves are familiar are exceedingly unstable and inconstant in the physico-chemical sense. Environments like these can only be inhabited by animals which, like ourselves, possess devices which allow changes in humidity, temperature and so forth to take place externally without producing corresponding internal changes in the inhabitants. In many ways the terrestrial type of environment is advantageous however, for terrestrial conditions are especially favourable for the growth of plants, which means an abundance of food for herbivorous animals and these, in their turn, can feed large numbers of carnivores. Again, an abundant supply of oxygen is available in the atmosphere, but abundance of food and oxygen are only two of the many factors needed if the dry land is to be a paradise for animals in general. These self-same terrestrial conditions also include periods of intense cold and intense heat, and may vary between drought at one time and extreme humidity if not positive flooding at others, not to mention tornados and hurricanes.

Very much greater physico-chemical stability is to be found in any large body of water. Because of the high viscosity and large specific heat of water, aquatic environments—except where the water is shallow—are relatively free from the wide and often violent thermal and mechanical disturbances met with on the land. Moreover, the free availability of water is a matter of the utmost importance for living things of every kind, because water is their chief constituent. As J. B. S. Haldane once remarked, even the Archbishop of Canterbury is 65 % water. Furthermore, water plays a larger or smaller part in most of the chemical reactions that take place in living organisms and upon which the very life of those organisms depends.

Because of the greater inherent constancy of water, especially when the volume is large, it seems likely that the need for complicated regulatory mechanisms must be less for aquatic than it is for landliving animals. Moreover, because of its enormous bulk and consequent physico-chemical inertia, the sea is probably the only environment that could provide the large degree of physico-chemical constancy which must have been needed to maintain the life of early, primitive and probably delicate living systems which as yet had devised no machinery for combating external changes and climatic conditions.

INTERNAL ENVIRONMENT: IONIC BALANCE

What has been said so far is pretty largely a matter of guesswork, for these early and primitive creatures have long since disappeared,

3

so that whatever we may say or think about their habits, habitats and behaviour must necessarily be a matter mainly of conjecture. We have to try to deduce the nature and properties of these long-vanished organisms from those of the creatures that survived, evolved and are still with us today. But most biologists would agree that the ocean was the most probable cradle of the earliest ancestors from which we all eventually descended.

However, these conjectural hypotheses are not the only arguments we can produce to support the idea of an oceanic origin of life. So far we have considered the external environment for the most part. Now it happens that physiologists, and biochemists too for that matter, often want to study parts of animals and therefore have to be able to keep the parts alive in isolation from the animal itself, and the ability to do this has some important theoretical consequences. All laboratory experience of this kind goes to show that, if the right conditions are provided, the heart or the liver of an animal, even a warm-blooded animal, can be kept alive and functionally normal for longer or shorter periods in complete isolation from the animal from which it came. The author, to give an example, has seen the heart of a snail still beating in a perfectly normal manner no less than a fortnight after removal from its former owner.

Because it is only by studying isolated organs and tissues that we can solve many of our problems, physiologists and biochemists owe what is probably their greatest single debt to Sidney Ringer. It was Ringer who was the first to show, in the early 1880's, that the isolated heart of a frog or a tortoise will survive and continue to beat normally for long periods in a solution containing only the chlorides of Na^+, K^+ and Ca^{2+}. Legend has it that Ringer worked for many months with only NaCl and KCl in different proportions, without finding any satisfactory mixture until one day his laboratory assistant inadvertently made up the solutions with tap water instead of distilled water, thus introducing calcium ions. Without these calcium ions life cannot be maintained. It is necessary only, as Ringer then showed, that the three chlorides should be present in the right proportions and that the total amount of salts should be about right, i.e.

Na^+	K^+	Ca^{2+}	Total
100	1·7	1·0	0·116M

With slight modification and the addition of small amounts of Mg^{2+} these simple solutions can maintain the heart beat of even a warm-blooded animal, but in this case the fluid must be kept neutral, well oxygenated and at the animal's own body temperature. One such solution devised by Tyrode has the following composition:

Na^+	K^+	Ca^{2+}	Mg^{2+}	Total
100	1·8	1·2	0·7	0·155 M

All this is interesting and important enough if only as a practical matter, but it has great theoretical importance as well. If we compare the compositions of the bloods of different kinds of animals we get two surprises. First, the proportions of ions present in different bloods are always very similar to each other and to Ringer's and Tyrode's solutions. Secondly, and perhaps more surprising still, the relative proportions of Na^+, K^+ and Ca^{2+} are very similar indeed in animal bloods and in sea water (see Table 1).

Table 1. *Relative ionic composition of animal tissue fluids*
(The amount of the commonest cation (Na^+) is taken as 100 to facilitate comparison.)

Animal	Na^+	K^+	Ca^{2+}
Sea water (approx.)	100	2·16	2·27
Coelenterates: *Aurelia* (jelly fish)	100	2·89	2·14
Annelids: *Amphitrite*	100	3·20	2·33
Molluscs: *Eledone* (octopus)	100	2·86	2·72
Mytilus (edible mussel)	100	2·52	2·51
Arthropods: *Carcinus* (shore crab)	100	2·32	2·51
Cancer (edible crab)	100	2·22	2·60
Homarus (lobster)	100	2·02	3·76
Astacus (freshwater crayfish)	100	1·84	7·9
Echinoderms: *Echinus* (sea urchin)	100	2·51	2·42
Asterias (star fish)	100	1·80	1·95
Fish: *Myxine* (hagfish)	100	1·72	2·24
Lampetra (lamprey)	100	2·67	3·27
Muraena (conger)	100	0·93	3·64
Lophius (angler fish)	100	2·80	1·01
Amphibia: *Frog*	100	2·41	1·92
Mammals: *Rat*	100	4·27	2·14
Man	100	3·51	1·72

Although there are large differences in *total* ionic composition and *total* salinity, the *relative proportions* of the most important cations are remarkably similar and certainly far more similar to each other and to those in sea water than would ever have been expected. This, surely, is evidence that life, as we know it today, must have originated in the ocean waters.

From the sea, some animals spread into the estuaries of rivers while others moved across the shore-line to the land, and a few, such as the land crabs and some terrestrial snails, reached dry land in that way. For the majority of animals however the route to land was the more circuitous one:

ocean → estuaries → fresh water → land.

Whatever the route, however, every move towards a new external environment confronted the animals with new structural and chemical problems which had to be solved before the move could be accomplished. To take only one example, animals accustomed to breathing in sea water, or in fresh water for that matter, could only live on the dry land when suitably modified respiratory organs—lungs instead of gills—had been developed, and when the blood itself had been so modified as to enable each species to take the oxygen it needed from air instead of from water. Any animal leaving the security of the relatively constant external environment provided by the sea would need new devices and mechanisms that would ensure internal constancy in the face of unfamiliar external variations.

In spite of their evolution and migrations into different habitats animals as a whole have maintained in their bloods and tissue fluids the old original relative ionic composition of an ancient sea water, and this has evidently been a matter not of choice but of obligation. Even the cells and organs of animals whose ancestors, like our own, became independent of the sea many millions of years ago, cannot tolerate for long any appreciable departure from the normal, sea-water-like composition of the blood as far as Na^+, K^+ and Ca^{2+} are concerned. This necessary internal constancy is something that *has* to be maintained. This is accomplished largely by the excretory organs; indeed, these organs must have been intimately concerned with the maintenance of internal ionic composition long before their functioning as excretory organs proper came into prominence.

TOTAL IONIC CONCENTRATION

Whereas the composition of the bloods of all animals is held constant as far as the *relative* proportions of Na^+, K^+ and Ca^{2+} are concerned, wide variations are found in the *total* ion contents and therefore in the osmotic pressure of the bloods of different animals. As has been mentioned already, the majority of animal migrations towards the dry land proceeded by way of the estuaries of rivers, into fresh water and thence to the land. Now in the estuaries large changes

Total Ionic Concentration

in total salinity take place with every tide. At high tide the water consists mainly of sea water, with a total salinity approaching 3 % of dissolved solids, while at low tide the water is mainly fresh, with a total dissolved solid content of only about 0·03 %.

In order to study the impact of these changes upon animals that have successfully penetrated into fresh water by way of estuaries, it is convenient to have some simple way of measuring and expressing *osmotic pressure*. Direct measurements present many problems. But the osmotic pressure of a solution is a measure of the number of dissolved particles per unit volume of the solution. Now the presence of dissolved matter not only confers osmotic properties upon the solution; it raises the boiling point, depresses the freezing point and, indeed, alters many of the physical properties of the pure solvent. All these alterations are proportional to the concentration of dissolved matter and any of them can therefore be used as a measure of the osmotic pressure, which is itself proportional to the concentration of dissolved particles, at any rate in fairly dilute solutions. It is usually convenient to determine the depression of the freezing point and to use this as a measure, indirect admittedly but very convenient, of the osmotic pressure of an aqueous solution. Thus sea water freezes at about -2 °C and fresh water at about $-0·02$ °C and this is usually expressed by the symbol Δ; thus for sea water $\Delta = -2$ °C and so on.

If we take a typical marine animal such as a spider crab or a sea urchin and place it in sea water diluted with increasing proportions of fresh water we find that the animal, which is accustomed to life in the sea, has no power of adapting to dilution of its customary external medium. Analysis of the bloods of animals kept under these conditions shows that salts are lost from the blood so that the internal osmotic pressure falls within a short time to a level equal to that of the surrounding water (see Fig. 1). If dilution is carried too far the animal dies. If however, we take an estuarine animal like the familiar little shore-crab (*Carcinus*) and repeat the experiment, it turns out that in this case, while dilution of the external medium is followed by a modest fall of internal osmotic pressure, the fall is less precipitous than it is in the spider crab (*Maia*) and never large enough to cause death, unless the animal is placed in mixtures containing only 1 % or 2 % of sea water. The results of some experiments of this kind are shown diagrammatically in Fig. 1. In *Carcinus* we have an animal that contrives to live well under estuarine conditions, but has never

7

so far contrived to push on into wholly fresh water. But there are other crabs and, for that matter, a considerable variety of other animals, that can stand up to completely fresh water, and not a few of them spend the greater part of their time in fresh water and return

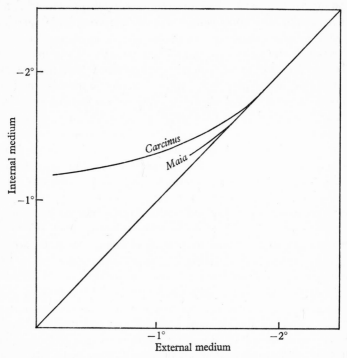

Fig. 1. Influence of dilution of external medium on internal medium in *Carcinus* and *Maia*. Osmotic pressure in terms of depression of freezing point. See text for explanation.

to the sea only at the breeding season. In such animals as these, curves somewhat similar to those found with *Carcinus* are obtained; usually transference to fresh water results in a slight dilution of the internal medium, but never large enough to be fatal. Well-known examples among the fishes are the common eel, which goes out to the Sargasso Sea in order to spawn, and the salmon, normally an inhabitant of the sea, which comes into the fresh waters of the rivers at the spawning season. The osmotic behaviour of these fishes has been studied, and results for the eel are shown in Fig. 2.

Total Ionic Concentration

It will be noticed in Fig. 2 that the internal osmotic pressure of the eel is never as high as that of sea water, and never as low as that of fresh water. This is generally true too of animals, which, having once overcome the special osmotic and other difficulties of estuarine existence, have settled down to a permanent life in fresh water. But, whereas those that are still estuarine, or traverse the estuaries at the breeding season, can still tolerate wide external variations in osmotic

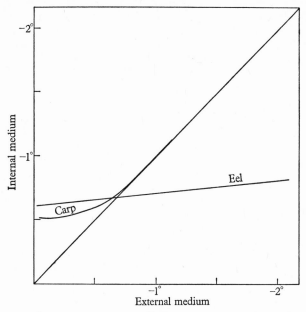

Fig. 2. Effect of changes in external medium on internal medium. The eel can tolerate transference from fresh to salt water and vice versa. The carp has little tolerance to external changes in comparison with the eel. Osmotic pressure in terms of depression of freezing point.

pressure, those that have settled down and remain permanently in fresh water have lost the ability to tolerate large external variations. To exemplify this, Fig. 2 shows the contrast between the eel and the freshwater carp. Just as, if we put a wholly marine animal into diluted sea water it dies if the dilution is too great, so too a wholly freshwater animal put into 'concentrated' fresh water dies if the osmotic pressure of the external water is too high.

So, then, marine animals rely upon the surrounding sea water for the maintenance of an internal medium of constant osmotic pressure, and freshwater animals similarly rely upon the surrounding fresh water. Animals that live habitually in estuaries or migrate from sea to fresh water from time to time experience and can tolerate wide variations in external osmotic pressure. They survive because they have the power of remaining osmotically independent of these external changes. Osmotic independence involves a number of mechanisms too numerous and too complicated to discuss here at any length, but the case serves well to illustrate how animals coming up against new and variable conditions in their external world have become adapted by the development of devices for controlling internal variations and can survive under conditions which, in the absence of the appropriate controlling mechanisms, prove fatal.

2

RESPIRATORY FUNCTION OF THE BLOOD

TRANSPORT OF OXYGEN

We have so far been thinking of blood as a medium which provides water and the necessary simple inorganic ions in the right proportions and at the right concentrations for the survival of cells and tissues. The discussion in the last chapter has shown that although the tissues can become adapted to work at rather different levels of total salinity, they have never succeeded in tolerating appreciable changes in ionic balance. In the vast majority of animals, however, the blood has many further functions to discharge.

With only a few exceptions, animals use up oxygen and produce carbon dioxide, and all but the most primitive possess some sort of blood, a heart and a circulatory system of some kind. As it passes through the specialized respiratory organs, the gills or lungs as the case may be, the blood takes up oxygen from the surrounding water or air and sheds its load of carbon dioxide. As it passes then through the tissues it hands over oxygen to those that require it and takes over the carbon dioxide that the tissues have produced.

There are quite a few animals whose blood still consists of virtually nothing but more or less diluted sea water, and they are usually comparatively sluggish, slow-moving or sedentary creatures with correspondingly small oxygen requirements. Plain sea water can accommodate about 0·5 ml. of dissolved oxygen per 100 ml., and this suffices to fulfil the oxygen requirements of animals of this kind.

The development of greater activity, and above all the much later development of warm-bloodedness necessitated, and was accompanied by, the development of more complex circulatory apparatus and more efficient respiratory organs. But these modifications alone did not suffice. In most animals, changes have taken place in the blood itself, *chemical* changes which make possible the uptake and transport of more oxygen per unit volume of blood. One such change consisted in many animals, including all the vertebrates, in the development of the familiar red pigment, *haemoglobin*, and its inclusion in the blood.

Haemoglobin, the commonest respiratory pigment known, has the ability to take up a relatively large volume of oxygen when exposed to a medium in which the partial pressure of oxygen is high, and to part readily with it again when the partial pressure is low. Haemoglobin is therefore ideally suited for the uptake of oxygen in the respiratory organs of an animal and for handing over that oxygen to the tissues.

Haemoglobin is an iron-containing pigment, and each of its Fe^{2+} ions can combine with 1 molecule of oxygen. It is present not only in the bloods of vertebrate animals but in many invertebrates too; in fact of the four known respiratory pigments, haemoglobin is far and away the commonest and most efficient. Nevertheless, other substances have been developed and used for oxygen transport in some groups of animals, but they are usually confined to particular groups of animals instead, like haemoglobin, of turning up haphazard all over the animal kingdom. Two large groups of invertebrate animals, the molluscs and the arthropods, possess a respiratory pigment that is peculiarly their own, this time a blue coloured, copper-containing substance known as *haemocyanin*. Two other such pigments are *haemerythrin*, which is reddish in colour and relatively uncommon, and the green *chlorocruorin*, which is confined to a small group of marine worms. Apart from haemocyanin all of these substances contain iron and all do essentially the same thing; they all combine freely and reversibly with oxygen, and they operate as agents which serve to transport oxygen from the respiratory organs to the tissues and to increase the oxygen capacity of the blood at the same time. Some important features of these four substances are summarized in Table 2.

Table 2

	Haemoglobin	Chlorocruorin	Haemocyanin	Haemerythrin
Colour	Red	Green	Blue	Red
Metal	Fe	Fe	Cu	Fe
Prosthetic group	Haem	Haem	Polypeptide	?
Molecule oxygen per atom metal	1:1	1:1	1:2	1:3
General properties	Two sharp spectral bands	Two sharp spectral bands	No sharp spectral bands	No sharp spectral bands
Occurrence	Corpuscles or plasma	Plasma	Plasma	Corpuscles or plasma

Transport of Oxygen

The most important common property of these pigments is the *reversibility of their reaction with oxygen.* Pyrogallol, as is well known, combines readily with oxygen, but would be useless as a respiratory pigment because it will not part with the oxygen again.

In its simplest form the behaviour of a respiratory pigment can be expressed by the equation

$$\text{Hb} + \text{O}_2 \quad \underset{\text{tissues}}{\overset{\text{respiratory organs}}{\rightleftharpoons}} \quad \text{HbO}_2$$

(Hb represents haemoglobin)

and by the curve (known as the dissociation curve) shown in Fig. 3.

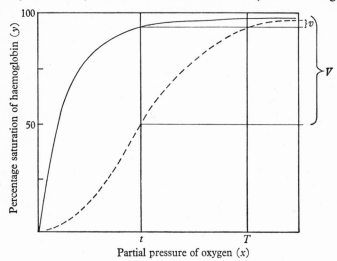

Fig. 3. Dissociation curves of haemoglobins. T = loading tension; t = unloading tension in tissues.

With regard to the equation one point must be emphasized. The equation refers to what is known as the *oxygenation* of haemoglobin; the process is not one of oxidation, for the Fe^{2+} ion of the haemoglobin is present in the ferrous form in both Hb and HbO_2.

By applying quite simple mathematics to this equation we can arrive at a general equation that fits the whole family of curves of which those shown in Fig. 3 are two rather extreme examples; that equation is,

$$y = 100 \; \frac{kx^n}{1 + kx^n},$$

where y = percentage oxygenation of the pigment, x = partial pressure of oxygen, k = dissociation constant of the oxyhaemoglobin,

13

and n is a constant the value of which varies from one blood to another. In the particular case of the full line in Fig. 3, n is unity. Sometimes the curve is a section of a simple rectangular hyperbola, like the full line in the figure, but in most cases, and especially among the higher and more active animals, the curve is more or less S-shaped, like the broken line in Fig. 3. The S-shapedness is a result of several factors, including the number of Fe^{2+} ions per molecule of haemoglobin, the concentration of inorganic salts and carbon dioxide, and the temperature. The degree of S-shapedness finds its expression in terms of the value of the constant n. Let us consider the simpler case first.

Suppose that the partial pressure of oxygen at the respiratory organs is T and that in the tissues it averages t, then the load of oxygen shed by the blood as it passes through the tissues is that shown as v in the diagram. If, however, the blood is of a kind in which n is greater than unity and the curve is pronouncedly S-shaped, a much larger part (V) of the oxygen available will be shed for the same fall in partial pressure. The constant n, which numerically expresses the degree of S-shapedness, is thus a measure also of the efficiency of the blood.

At least two other factors are concerned in determining the efficiency of the blood, first the nature and oxygen-combining power of the pigment, and secondly, the concentration of pigment in the blood. As for the first point, look again at Table 2. From this point of view, atom for atom of metal, chlorocruorin is as efficient as haemoglobin, but has been hit upon by only a few worms. Haemocyanin is half as efficient and haemerythrin only one-third as efficient as haemoglobin.

Table 3. *Oxygen capacities of some different bloods*

Pigment	Colour	Site	Animal	ml oxygen per 100 ml blood
Haemoglobin	Red	Corpuscles	Mammals	25
			Birds	18·5
			Reptiles	9
			Amphibians	12
			Fishes	9
		Plasma	Annelids	6·5
			Molluscs	1·5
Haemocyanin	Blue	Plasma	Molluscs:	
			Gastropods	2
			Cephalopods	8
			Crustaceans	3
Chlorocruorin	Green	Plasma	Annelids	9
Haemerythrin	Red	Corpuscles	Annelids	2
Sea water	—	—	—	c. 0·5

Transport of Oxygen

With regard to the second point (Table 3), the concentration of pigment per unit volume of blood, haemoglobin is again the most efficient, for while haemocyanin and chlorocruorin are always and only found in solution in the blood plasma, haemoglobin is sometimes found in solution. But most usually it is present in special cells, the red blood corpuscles, which are literally packed with haemoglobin at very high concentrations. Red corpuscles occur in all vertebrate bloods, including our own, which can carry about one-quarter of its own volume of oxygen. Even higher oxygen capacities are known; in seals and other diving mammals, for example, the capacity may be as much as 50 % by volume, but here the haemoglobin serves an additional function. First of all it serves to transport oxygen from lungs to tissues, as it does in ourselves, but it serves too as a store of oxygen upon which the animal can draw while it is submerged.

There are, then, several stages in the evolution of an efficient blood, which may be summarized as follows:

(i) Sea-water-like blood.

(ii) Invention and incorporation of a respiratory pigment.

(iii) Increased S-shapedness of the dissociation curve.

(iv) Increased concentration of pigment.

(v) Invention of the red blood corpuscle.

Needless to say, different animals have different oxygen requirements, requirements dictated by the conditions under which they live, and their bloods are usually nicely adjusted to fulfil the particular requirements. It has, in fact, been shown that every animal species has its own specific respiratory pigment; no two haemoglobins are exactly the same. Rat, rabbit, pig and human haemoglobins are all different, and crab and lobster haemocyanins are different again. All of these pigments belong to the class of *conjugated proteins*, so called because they are formed from a protein with which is united a non-protein substance or *prosthetic group*. It is this non-protein prosthetic group that is the business part of the pigment, the part that combines with the oxygen. The differences between different haemoglobins are due to differences not in the prosthetic group, which is always the same, but in the protein component, which has accordingly been called the 'fine adjustment' of the haemoglobin molecule.

So much then for the transport of oxygen. Fortunately there is only a minute change in the reaction of the blood when it loads up with oxygen; actually there is a very slight increase in acidity, but the change is usually too small to have any very serious consequences for the cells and the tissues.

Respiratory Function of the Blood

The transport of carbon dioxide is another matter because this gas reacts slowly with water to form carbonic acid, which rapidly ionizes to give bicarbonate ions together with hydrogen ions, which make for increased acidity of the blood.

$$CO_2 + H_2O \rightleftharpoons H_2CO_3 \rightleftharpoons H^+ + HCO_3^-.$$
$$\text{(slow)} \qquad \text{(fast)}$$

The reactions can be written as reversible because they proceed in the forward direction when the blood loads up with CO_2 in the tissues and in the reverse direction when CO_2 is discharged at the respiratory organs. Here, therefore, there is a real and acute danger of significant changes in the acidity or alkalinity of the blood.

However, a certain amount of CO_2 can be transported *in simple solution*, and it is estimated that about 5 % of the CO_2 transported in human blood is dealt with in this way. A second method of transport depends upon a reaction with the proteins of the blood plasma. We shall have a good deal more to say about these very complex substances later on, but for the moment it suffices to know that they carry a variety of groupings among which amino-groups, $-NH_2$, and carboxyl groups, $-COOH$, are common. The $-NH_2$ groupings are capable of reacting with CO_2 to form what are known as *carbamino-compounds*:

$$(Pr) - NH_2 + CO_2 \underset{\text{lungs}}{\overset{\text{tissues}}{\rightleftharpoons}} (Pr) - NH.COOH.$$

This mechanism is thought to be responsible for about 10 % of the CO_2 transported in human blood. But even when solution and carbamino-compound formation are taken together there remains another 80–85 % of CO_2 to be dealt with.

In comparatively slow-moving creatures with small oxygen-requirements the problem of CO_2 transport is correspondingly small. All animal bloods can carry some CO_2 in solution, and the fact that its reaction with water is relatively slow minimizes the rate at which hydrogen ions are produced. Moreover, even comparatively simple, sea-water-like bloods usually contain more or less protein, so that carbamino-compound formation can be called in as a second means of CO_2 transport. But in more active and highly organized animals things are more complicated because the blood contains a zinc-containing catalyst, known as *carbonic anhydrase*, which accelerates the reaction between CO_2 and water by a factor of several hundred

16

times. In our own blood, and in the blood of mammals generally, this catalyst is present, so that hydrogen-ion formation is a rapid and a large-scale affair. The production of H^+ is tantamount to the formation of free acid; in fact acid concentration is usually measured and expressed in terms of H^+ concentration.

No living cell or tissue can tolerate changes in the concentration of H^+ for long or on any appreciable scale and we must inquire what becomes of these ions. In part, but only in small part, they are dealt with by reacting with the small amounts of phosphate ions present in the blood:

$$H^+ + HPO_4^{2-} \rightleftharpoons H_2PO_4^-.$$

A larger part is dealt with by the carboxyl groups of the blood plasma proteins (Pr), which under the conditions in which they occur in blood, are ionized and therefore negatively charged:

$$(Pr)\,COO^- + H^+ \rightleftharpoons (Pr)\,COOH.$$

Any dissolved respiratory pigment that happens to be present can play a part in the same way as the plasma proteins.

So far, then, we have discovered three ways in which CO_2 can be transported in the blood, namely:

(i) In simple physical solution.

(ii) As carbamino-compounds with the plasma proteins.

(iii) As HCO_3^-, the corresponding H^+ being held by the plasma proteins and, to a less extent, by the blood phosphates.

Any or all of these processes can be used and, of course, more CO_2 can be dealt with by bloods rich in protein than by those that contain little or none. One point of importance is worth mentioning before we go further. It concerns the oxygen-transporting respiratory pigments, which are themselves proteins. Their presence in the blood serves not only to increase the supply of oxygen to the tissues; because they are proteins, they provide at the same time increased CO_2 capacity and increased H^+-combining power. High oxygen consumption means high CO_2 production, and *both are facilitated by one and the same adaptation*, i.e. the inclusion of a respiratory pigment in the blood.

As we saw earlier in this chapter, the highest level of respiratory efficiency as far as oxygen transport goes was realized with the invention of the red blood corpuscle. With this goes a remarkable and efficient mechanism for dealing with CO_2 and for accommodating H^+.

In corpusculated bloods like our own, carbonic anhydrase is present but *it is confined entirely to the inside of the red cells*. This means that little carbonic acid and few hydrogen ions are formed in the plasma, and such as are can be dealt with by phosphates and by the plasma proteins. The bulk of the CO_2 enters the red cells and there reacts rapidly with water under the catalytic influence of carbonic anhydrase. Carbonic acid is formed and ionizes spontaneously to give bicarbonate and hydrogen ions. These processes can be pictorially represented as in Fig. 4.

The red cells are tightly packed with haemoglobin, which is a protein and is negatively charged under the conditions prevailing

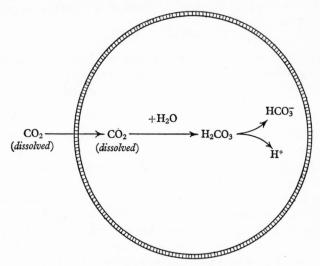

Fig. 4. Red blood corpuscle taking up a load of carbon dioxide.
For details see text.

inside the corpuscles, owing to the ionization of its carboxyl groups. This negatively charged haemoglobin takes up H^+ ions and converts them into what might be called 'haemoglobinic acid':

$$(Hb)^- + H^+ \rightleftharpoons H(Hb)$$

and it is in this way that 80–85 % of the CO_2 transport in human blood is handled. The CO_2 is carried as HCO_3^- and the corresponding H^+ are accommodated on the haemoglobin within the red cells.

While all these changes are going on, bicarbonate ions accumulate within the red corpuscles, and tend, therefore, to diffuse away into the plasma. If this happened, however, the departure of each negatively charged bicarbonate ion would leave one unit of positive charge behind. Every corpuscle would repel all the other corpuscles and the consequences would quite certainly be disastrous. It might be expected therefore that, for each bicarbonate ion leaving the cell, one positively charged ion would go along with it, thus preserving electrostatic neutrality. This might be done by the outward diffusion of potassium, the commonest cation in the red cells. For some reason that is not very clear, however, the corpuscular membrane behaves as though it were completely impermeable to positively charged ions such as potassium, and a different mechanism is actually used.

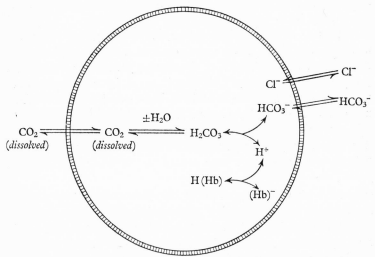

Fig. 5. Red blood corpuscle taking up or shedding a load of carbon dioxide. This illustrates the buffering effect of haemoglobin and the 'chloride shift'. For details see text.

The blood plasma is rich in chloride ions, and for each bicarbonate ion leaving the cell a chloride ion moves in. So that, although bicarbonate ions are formed within the red cells, they diffuse out and are carried mainly in the plasma, and for each molecule of CO_2 entering the cell, one hydrogen ion is formed and accommodated on haemoglobin inside the cell, while the bicarbonate ion leaves the cell and is replaced by a chloride ion. This phenomenon is usually referred to as the '*chloride shift*'. The whole process can be summarized by the revised version of Fig. 4 shown in Fig. 5; the reactions go forward when CO_2 is taken up from the tissues and in reverse when CO_2 is discharged at the respiratory organ.

By the mechanisms we have just considered, the blood of animals is maintained in a state very closely approximating to neutrality—

another example of the maintenance of constancy in the internal environment. One point that is worth special comment is that the red blood cells can be thought of as a respiratory tissue; not a fixed tissue like the liver or the kidney, but one that circulates. It is particularly significant that the large-scale production of H^+ that accompanies CO_2 uptake does not take place in the internal medium proper, as represented by the blood plasma, but is kept away from the internal medium because carbonic anhydrase is present only inside, and not at all outside, the cells of this mobile and circulating respiratory tissue.

Constancy of relative ionic composition, osmotic pressure and hydrogen-ion content of the blood are three examples of the numerous internal constancies which characterize animals as a whole. A good deal is known about the ways in which these constancies are maintained, and it is time to find out, if we can, not the *how* of the matter but the *why*. The answer is almost certainly this.

If any group of substances is more characteristic of living stuff than any other, pride of place must certainly go to the *proteins*. Not only do they make up a large part of the structure of living cells and tissues; they provide also the enormous army of catalysts, the enzymes, which control and direct the innumerable chemical reactions that underlie every activity of the living cell. Whether they play a structural part in a given cell or whether they act in a catalytic capacity, proteins are exceedingly sensitive to small changes in the physico-chemical properties of their immediate surroundings, and it is therefore only by the maintenance of these properties at constant levels that the structural and catalytic integrity—that is to say the very life—of the cells can continue.

3

PROTEINS

THE DIVERSITY OF PROTEINS

Since proteins make up a very large part of the structure of every cell and, moreover, provide the enzymes that control and regulate the whole of the cell's metabolism, it seems reasonable to consider the proteins before the other main tissue constituents, the carbohydrates and fats. These are of no less importance to the life of the cell as a whole, but a start has to be made somewhere.

Proteins are very numerous. Human blood plasma contains more than a score of different proteins, each with its own particular function. The blood of a rabbit contains a similar number of proteins, and although they resemble those of a man pretty closely they are not identical. Unfortunately, because of their extreme chemical complexity, it is possible in only a few cases at the present time to distinguish between man-proteins and rabbit-proteins by chemical methods, but there are other ways. If, for example, an animal is given a minute dose of a foreign protein by injection, nothing very much happens. If a rabbit is given a minute does of hen's egg albumin by injection, nothing very obvious happens, but if after the lapse of a week or two a second dose is given, the animal becomes sick and quickly dies of what is called *anaphylactic shock*. It has, in fact been *sensitized* to the foreign protein. But if the second dose of hen-protein is replaced by albumin from a duck's egg, the reaction is far less severe, a clear indication that the two proteins, similar though they are in appearance and chemical make-up, are not altogether identical. Incidentally many food idiosyncrasies and a number of other disorders originate in sensitization phenomena; hay fever is commonly due to sensitization to plant pollens; many persons cannot tolerate hen's eggs but get on successfully with duck's. How these sensitizations take place is not always at all clear, and one cannot help feeling sorry for people who react to strawberries and cream by a vicious nettle rash. However, if repeated injections of a foreign protein are given at short successive intervals there is no such violent reaction. Instead the animal becomes *immunized* against the unfamiliar protein, and this process of immunization has many important clinical applications. The reader has probably been immunized against smallpox, poliomyelitis and a few more diseases.

Even the haemoglobins of different animals are not identical but differ slightly in their affinity for oxygen, in their solubility and even in their crystalline form. Every species of animal, in fact, contains probably many hundreds of different proteins, and there are many thousands of different species of animals, each with its own set of proteins. So the total number of proteins in existence must, indeed be extremely great. How can this immense variety be produced?

It has been shown by analysis of the products of hydrolysis that proteins are made up of amino acids. In all about twenty such amino acids have been found in this way. Not all proteins contain all of these amino acids, but if we allow about 500 amino acid units as an average number for the formation of one protein molecule, and if we suppose that these different units can be chosen in any proportions and arranged in any sort of order within the protein molecule, then the number of theoretically possible proteins is incomparably greater than the number of visible stars in the sky. So there are plenty of opportunities for variation.

THE STRUCTURE OF PROTEINS

Every one of the amino acids that enters into the composition of proteins possesses at least one amino and one carboxyl group, and both are attached to the same carbon atom. Their general structure is therefore

$$R . CH \begin{cases} NH_2 \\ COOH \end{cases}$$

R represents different kinds of groupings (see Table 4, p. 27). Many of the amino acids give *special colour reactions* by which they can be recognized, whether free or in combination, and they all give a colour reaction with a reagent known as ninhydrin.

From the point of view of protein structure, however, the most important feature of the amino acids is that they can react one with another, the carboxyl group of one uniting by condensation with the amino-group of the next and so on to form a long, more or less coiled or spiralled string of amino acids joined by peptide linkages.

The Structure of Proteins

Here R_1, etc, correspond to the R groups of different amino acids. This type of peptide linkage is present in proteins of every kind. Any substance that contains two or more of these peptide linkages will give the so-called *biuret reaction*, so that the biuret reaction itself can be used as a general test for proteins.

This reaction gets its name from biuret, a simple compound in which two peptide links are present. It can be obtained by heating urea very gently so that ammonia is driven off and biuret remains:

$$\begin{array}{c} \underset{CO}{\overset{NH_2}{\diagup}} \\ \diagdown \\ NH_2 \\ + \\ NH_2 \\ \diagup \\ CO \\ \diagdown \\ NH_2 \\ \text{\textit{urea}} \end{array} \rightarrow NH_3 + \begin{array}{c} NH_2 \\ | \\ CO \\ | \\ NH \\ | \\ CO \\ | \\ NH_2 \\ \text{\textit{biuret}} \end{array}$$

In the presence of strong alkali and traces of copper ions, biuret and the proteins give a pale rose, pink or mauve coloration which varies somewhat from one substance to another, but is quite characteristic of proteins as a whole.

The other usual colour tests for proteins are not general tests but depend upon the presence of particular R groups in the molecule. Most of the amino acids are present in most proteins, but there do exist proteins which lack some particular amino acid or amino acids altogether. *Gelatin* is such a protein.

NUTRITIONAL IMPORTANCE OF PROTEINS

If you give a meal of protein to a dog—lean raw beef will do admirably—a quantity of nitrogen equivalent to that in the protein of the meal will appear within about twenty-four hours in the animal's urine in the form of *urea*. This shows that *the animal body cannot store proteins*. Fat and carbohydrate, the former sometimes in embarrassingly large quantities, certainly can be stored, but not protein.

However, if a dog is starved of protein for some days or weeks, urea excretion drops to a very low level, but it does not cease altogether. Even during starvation, proteins and amino acids are required for a number of special jobs in the body and, in the absence of incoming food protein, the animal has to break down some of its own tissue proteins to get the materials it needs. If this is allowed to go on happening, a time will come when the animal has used up all its reserve stores of fat and carbohydrate for energy production and has to fall back upon the proteins of its own tissues. These now begin to be broken down on a large scale, urea production rapidly increases to a high level, and shortly after this so-called '*pre-mortal rise*' in urea excretion, the animal dies.

To go back to the early part of this rather unpleasant experiment it will be realized that, although the protein-starved animal is not receiving any nitrogen in the form of protein, it is nevertheless losing small amounts of nitrogen in the form of urea, and this nitrogen comes from the breakdown of some of its own protein. Unless protein is eaten the animal will continue to lose nitrogen and eventually it will die. *Protein, therefore is an indispensable food.* Provided that enough protein is available, however, an animal can live on diets containing little or no fat or little or no carbohydrate for long periods. But it must have protein, and enough of it. How much is 'enough?'

Suppose that you take your protein-starved dog and feed it gradually increasing amounts of protein day by day, starting at a very low level. A time will come when the amount of protein nitrogen going into the animal just suffices to balance the amount of urea nitrogen coming out. The animal is then said to be in *nitrogen balance* and the amount of protein given corresponds to its *minimum protein requirement*. This is a matter of much interest, especially in times of war or famine. The minimum protein requirement of an average man is known with fair accuracy—we shall have more to say about this presently—and when rationing schemes have to be devised and put into action it is of the utmost importance that every man, women and child shall get his or her minimum protein requirement, or as much of it as conditions allow. If your experimental animal, man, woman or dog, gets less than this minimum the organism, is, on balance, the loser and will have to draw on its own tissues to make good the deficit. If it receives more than this indispensable minimum, the excess will merely be degraded and the nitrogen excreted as urea, since *protein cannot be stored* against a meatless day.

To this general rule about nitrogen retention there are some special exceptions. In young, growing animals, new tissue protein is being laid down and some protein nitrogen is retained to provide this. The same is true in pregnancy, when new baby-protein is being laid down in the foetus, and in lactation, when milk protein is being produced. The same is true during convalescence after a wasting

Nutritional Importance of Proteins

illness. So that in children, in pregnant and nursing mothers, and in convalescing patients the protein requirement per unit of body weight is higher than in normal healthy adults.

BIOLOGICAL VALUE OF PROTEINS

In spite of the nutritional importance of proteins for animals in general, the ability of animals to synthesize the constituent amino acids for themselves is very restricted indeed. Of the twenty-odd species of amino acids needed for the manufacture of animal proteins about one half cannot be synthesized by the animals themselves; at any rate they cannot be synthesized fast enough to keep pace with the demands of optimal growth. For all of these, young animals are entirely dependent on their food.

The green plants possess much more synthetic ingenuity. Unlike animals, they can use water, carbon dioxide and simple inorganic substances such as nitrate as their sole source of nitrogen and from these they can synthesize all the amino acids they need for development, growth and reproduction. Indeed, among bacteria there are some that can utilize even gaseous nitrogen from the atmosphere. In the end, therefore, animals have to rely upon the green plants. Herbivores get the amino acids they need at first hand by eating plants, and carnivores get them at second hand by eating either the herbivores or carnivores that are smaller than they are themselves.

Because of the nutritional importance of proteins attempts have been made from time to time to determine the minimum protein requirement of the average man. Apart from rationing schemes and the like, proteins are among the most costly articles of food, and although they may not be very interested in it, a knowledge of minimum requirements is important for families living on restricted budgets.

According to the experiments of Rubner, one of the leading German nutritionists in the early years of this century, the average man needs at least 100–125 g of protein food daily. Another worker, Chittenden, who used himself as the experimental animal, found that he could get by with only about 30 g a day, and, moreover, his health improved during the period of the experiments. We know now, however, that the large difference between these two estimates does not mean that different people have greatly different requirements. The reasons are more interesting than just this.

Food proteins are not absorbed by an animal as such. For one thing, cow-protein or sheep-protein is not the same as man-protein, and if a man is to rely on the cow or the sheep to keep up his own

25

levels of man-protein, the protein he eats has to be completely dismantled by his digestive enzymes, and the component amino acids must then be rearranged by the man's own enzymic apparatus to produce his own kind of protein. So that, in effect, the job of protein foodstuffs is really that of supplying enough of the right amino acids.

Now of the twenty-odd species of amino acids that go to the formation of the average protein the animal can, if need be, manufacture about a dozen for itself. The rest it cannot make fast enough to sustain a normal rate of protein production, especially in growing animals. For the balance, therefore, it is absolutely dependent upon its food. Consequently we can divide the amino acids into two groups, those which are *essential* and those which are *non-essential*, and the real job of protein food is to supply enough of those that are essential. In the end, of course, all these come from green plants.

As examples of essential amino acids we may mention *phenylalanine*, *tryptophan* and *lysine*, and, of the non-essentials, *glycine*, *alanine* and *glutamic acid* (for formulae see Table 4). It follows that no animal could live on a diet that contains no tryptophan or no lysine, no matter how much protein it might consume, though it could survive well enough on a diet containing no glycine, alanine or glutamic acid. This is a very important matter because some proteins are altogether lacking in particular amino acids and we have in fact already come across the case of *gelatin* (p. 23), a protein that contains no tyrosine and no tryptophan. However, gelatin is not by any means the only nutritionally deficient protein known.

The most famous protein of this kind, which occurs in maize and is called *zein*, contains no tryptophan and no lysine. Two American workers, Osborne and Mendel, fed young rats on a diet in which zein was the sole protein. The animals stopped growing almost at once, began to lose weight and were well on the way to the grave when the experiment ended. Next they were given the two missing amino acids in addition to their ration of zein and soon resumed their weight and rate of growth. Others experiments were done in which only lysine was given, together with the zein. In this case the animals did not lose any weight but, like Peter Pan, they refused to grow up, and it was only when tryptophan also was given that they began to grow again.

By and large it is true to say that vegetable proteins, such as zein, are less rich in essential amino acids than are proteins of animal origin so that, if you eat nothing but plant proteins, you will need

Biological Value of Proteins

to eat a lot more total protein than if you go in for meat, fish, eggs, milk or cheese. And this is the explanation for the discrepancy between Rubner's results and those of Chittenden (p. 25). Chittenden used nothing but animal proteins in his experiments, whereas Rubner used some animal protein and a large proportion of plant protein. Here is a good case where the experts disagree and yet are both right at the same time.

Table 4. *Formulae of some important amino acids*

Name	Formula	Text reference on page
Glycine	$CH_2(NH_2)COOH$	26, 32–5, 82
Alanine	$CH_3CH(NH_2)COOH$	26
Serine	CH_2OH | $CH.NH_2$ | $COOH$	—
*Cysteine	CH_2SH | $CH.NH_2$ | $COOH$	—
*Valine	CH_3CH_3 \\ / CH | $CH.NH_2$ | $COOH$	—
Arginine	$HN=C$ NH_2 / \\ NH | $(CH_2)_3$ | $CH.NH_2$ | $COOH$	35, 51–6
*Lysine	NH_2 | $(CH_2)_4$ | $CH.NH_2$ | $COOH$	26, 35

27

Table 4 (*cont.*)

Name	Formula	Text reference on page
*Methionine	$S.CH_3$ | CH_2 | CH_2 | $CH.NH_2$ | $COOH$	—
Glutamic acid	$COOH$ | CH_2 | CH_2 | $CH.NH_2$ | $COOH$	26, 35, 49

Proline

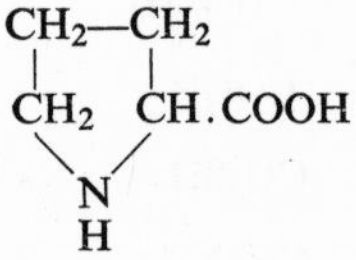

$$CH_2\!-\!CH_2$$
$$CH_2 \quad CH.COOH$$
$$\diagdown N \diagup$$
$$H$$

—

*Tryptophan $CH_2CH(NH_2)COOH$ 26

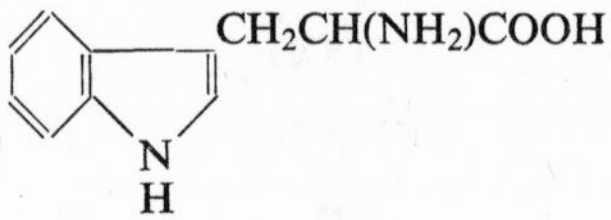

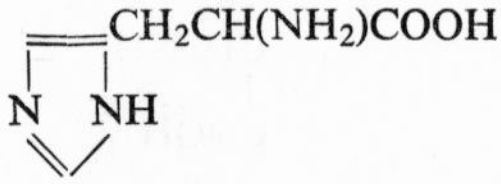

*Histidine $CH_2CH(NH_2)COOH$ 26

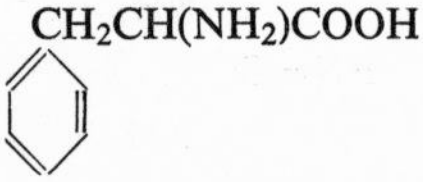

N NH

*Phenylalanine $CH_2CH(NH_2)COOH$ 26, 106–8

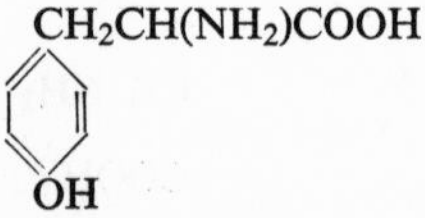

*Tyrosine $CH_2CH(NH_2)COOH$ 106–7

OH

Notes. (1) An asterisk indicates that the amino acid is '*essential*', i.e. cannot be synthesized by the animal for itself; either too slowly to keep pace with normal growth, or in some cases not at all. (2) *Glycine* is essential for young chicks. (3) *Arginine* is advantageous but not absolutely essential in young mammals (4) *Cysteine* is replaceable by methionine, from which it can be formed. (5) *Tyrosine* is replaceable by phenylalanine, from which it can be formed. (6) The page numbers correspond to references in the text.

4

PHYSICO-CHEMICAL BEHAVIOUR
OF PROTEINS

GENERAL PHYSICO-CHEMICAL PROPERTIES

The properties of proteins, especially those of the biological catalysts we call enzymes, are profoundly affected by changes in the properties of the media in which they operate, and particularly by the acidity or alkalinity of those media. Before we can deal properly with the properties of proteins, therefore, a brief diversion is desirable. Proteins, chemically speaking, are very complex substances and are ampholytic, i.e. they can behave as weak acids or weak alkalis. Let us therefore recollect a few of the properties of weak acids.

By definition, acids are acidic because they give rise in aqueous solution to hydrogen ions:

$$HX \rightleftharpoons H^+ + X^-.$$

The character that determines the strength of an acid is the extent to which this ionization prevails. Weak acids are mainly un-ionized, and we can use a special though rather unorthodox notation to emphasize the properties of a weak acid:

$$HX \xrightarrow{\quad} H^+ + X^-.$$

This is not meant to imply that the rates of the forward and reverse reactions are unequal but only that the equilibrium lies strongly in favour of the direction indicated by the longer of the two arrows.

Let us for a moment consider the behaviour of a salt, say the sodium salt, of a weak acid, HX. Being a salt the compound will be almost completely ionized in aqueous solution:

$$NaX \xrightarrow{\quad} Na^+ + X^-.$$

Now suppose that we take a $N/100$ solution of a strong acid, say HCl, we know that it will ionize fully so that the hydrogen-ion concentration will be approximately $N/100$. If, however, we add a solution of NaX the following reactions will take place:

$$\text{(i) } HCl \xrightarrow{\quad} H^+ + Cl^-,$$
$$\text{(ii) } NaX \xrightarrow{\quad} Na^+ + X^-,$$
$$\text{(iii) } H^+ + X^- \xrightarrow{\quad} HX.$$

The result will be that a large proportion of the hydrogen ions will be removed from the solution. In other words, the salt of the weak acid acts as a 'buffer'; it mops up hydrogen ions. We have already seen that the proteins of blood plasma and the haemoglobin and other respiratory pigments can act in this way. They all behave as exceedingly weak acids, are negatively charged under the conditions that prevail in the blood, and are capable of acting as acceptors of hydrogen ions (pp. 17, 18).

Since the properties of proteins are as profoundly affected as in fact they are by changes in acidity, it is desirable to have some convenient way of measuring acidity and of expressing it in a convenient, numerical, and therefore precise, manner. This can be done in terms of normality, 10^{-1} N (N/10), 10^{-2} N (N/100) and so on, but troubles arise when we attempt to plot a graph using these 'numbers' as abscissae. We get a co-ordinate like this, with all the points tending to bunch up towards one end of the scale:

To get over this it is possible to make our plot on logarithmic paper, when the points will fall with equal distances between them; but logarithmic paper is very expensive whereas tables of logarithms are cheap and can be used over and over again. Instead, therefore, of using logarithmic paper we can plot logarithms directly. If, to take a simple example, we have N/10 and N/100 acid, we can re-express these as 10^{-1} N and 10^{-2} N, from which we see at once that the logarithms are -1 and -2 respectively in the two cases. In the general case $[H^+] = 10^{-x}$, we could plot the values of $-x$, the logarithms of a series of figures. To make matters even more easy we can omit the negative sign and simply plot values of x. The only snag about this is the fact that as x *increases* the acidity beomes *less*, but we get a very convenient axis:

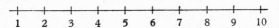

To avoid the necessity of constantly referring to 'minus the logarithm of x' we must have a name that represents it and this is the quantity spoken of as pH. Originally this was written as p_H, which caused typographers so many headaches that the simpler expression pH was eventually adopted all round. Many students seem to think there is something mysterious about pH. There is not. It is simply a matter of convenience and definition.

If $[H^+] = 10^{-x}$,

then $\log [H^+] = -x$,

therefore

$$x = pH = -\log [H^+].$$

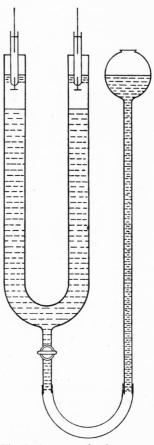

It is important to notice in passing that pure water is taken as the ideal standard of neutrality. Pure water is very slightly ionized, with $[H^+] = 10^{-7}N$; on the pH scale therefore neutrality corresponds to pH 7.

It has been pointed out already that proteins are very sensitive to changes in pH. One of the reasons for this can be demonstrated by means of the apparatus illustrated in Fig. 6. A buffer solution of known pH is run into the lower part of the U-tube. Then, very cautiously, a coloured protein solution of the same pH, a solution of haemoglobin for example, is run in from below so as to give a clean, sharp boundary on each limb of the tube. Electrodes are inserted at the top of each of the two limbs and a direct current is switched on. If this is repeated using a series of buffers each at a different pH it turns out that at pH's below about 6·8 the boundaries

Fig. 6. Apparatus for demonstration of electrophoresis. See text for explanation.

move towards the cathode. This phenomenon is called *electrophoresis* and it shows that the haemoglobin is positively charged. At higher pH's the boundaries move towards the anode, showing that the protein is negatively charged. At pH 6·8 the boundaries do not move, showing that the protein is electrostatically neutral.

This behaviour is typical of proteins in general. In every case there is some particular pH at which the protein does not move in an electric field, indicating that it carries no effective charge, and this is known as the *isoelectric pH* or isoelectric point of the protein in question.

A further phenomenon that is common to proteins in general is that many of their properties, e.g. solubility, osmotic pressure, viscosity, electrical conductivity and so on, and, in the case of enzyme proteins, their catalytic activity, all pass through either a maximum or a minimum at the isoelectric pH. One can imagine therefore that in a living cell, where a large number of proteins are working harmoniously together, a small change of pH in the cell might well increase the solubility of some and decrease that of others, leading to a state of disharmony and so to a breakdown in the normal physiological functioning of the cell as a whole. And solubility is only one of the many properties that would be similarly affected.

Proteins, then, carry different charges at different values of pH. From this it can be inferred that they carry numerous ionizable groups, some of which are ionized in solutions more acid than the isoelectric pH while others are ionized in more alkaline solutions. This, indeed, is true, and the ionizable groups take their origin in the building blocks from which proteins are constructed.

PHYSICO-CHEMICAL BEHAVIOUR OF AMINO ACIDS

These building blocks are *α-amino acids*, all of which can be represented by the general formula, $R.CH(NH_2)COOH$. In all about twenty such substances have been found by the hydrolysis of different kinds of proteins, and in all probability the two simplest are already known to the reader, namely, *glycine*, in which R corresponds to a hydrogen atom, and *alanine*, where R is a methyl group:

$$CH_2(NH_2)COOH \qquad CH_3.CH(NH_2)COOH$$

glycine   *alanine*

Not all proteins contain all of the twenty-odd amino acids, many of which give characteristic colour reactions when present as constituents of proteins so that it is a comparatively easy matter to find out whether or not egg albumin, for example, contains this or that particular amino acid.

Now the foregoing paragraph contains a serious error for, although amino acids in general are formulated as $R.CH(NH_2)COOH$, they do not normally exist as such in solution. Sodium chloride is similarly formulated as NaCl, though in solution it exists almost exclusively as its ions, Na^+ and Cl^-. Similarly, glycine, to take only the simplest example, does not exist in isoelectric solution (pH 6·1) as

$$\begin{array}{c} NH_2 \\ \diagup \\ CH_2 \\ \diagdown \\ COOH \end{array}$$

but in the fully ionized state,

$$\begin{array}{c} NH_3^+ \\ \diagup \\ CH_2 \\ \diagdown \\ COO^-. \end{array}$$

The pH of an amino acid in isoelectric solution may be anywhere between 3 and 10, depending on the nature and number of the ionizable groups in its 'R' group. This is equally true of alanine and all the rest of the amino acids, and the point can be proved by titrating with acid and with alkali. If the titration is repeated after treating the amino acid with formaldehyde, which reacts with and blocks the $-NH_3^+$ groups

$$-NH_3^+ + HCHO \rightarrow {}^-N:CH_2 + H_2O + H^+,$$

the amino acid no longer titrates with alkali, as is shown in Fig. 7. This shows that when we titrate with acid we are, in fact, titrating not the 'basic' amino group but the 'acidic' carboxyl group and that what is happening is as follows:

$$\begin{array}{ccc} NH_3^+ & NH_3^+ & NH_2 \\ \diagup \quad {+H^+} & \diagup \quad {-H^+} & \diagup \\ R.CH \quad \longleftarrow & R.CH \quad \longrightarrow & R.CH \\ \diagdown & \diagdown & \diagdown \\ COOH & COO^- & COO^- \\ acid & neutral & alkaline \end{array}$$

In other words, the addition of acid to an isoelectric solution of an amino acid does not, as was at one time believed, lead to the ionization of the amino group but to suppression of that of the carboxyl group. Similarly, the addition of alkali to an isoelectric solution of an amino acid leads to suppression of the ionization of the amino-group, and not to ionization of the carboxyl.

It follows from all this that there is a considerable resemblance between a simple amino acid like glycine or alanine on the one hand and a typical protein on the other. In both cases there is some one particular pH at which the molecule carries *no net effective charge*. This is not to say that the charge is zero, but that *the numbers of*

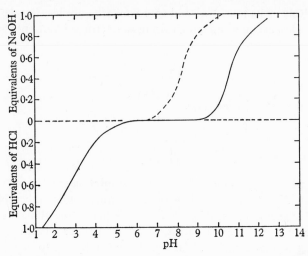

Fig. 7. Titration curves of glycine. The full curve corresponds to titration in aqueous solution and the broken line to that after treatment with formaldehyde. (After Fruton and Simmonds.)

Table 5. *Molecular weights of some proteins*

Protein	Source	Molecular weight
Insulin	Pancreas (ox)	5,734†
	(pig)	5,778†
	(sheep)	5,704†
Ribonuclease*	Pancreas	12,700†
Lysozyme	Egg-white	17,000
Pepsin*	Gastric juice	34,500
Egg albumin	Egg-white	44,000
α-Amylase*	Pancreas	45,000
	Malt	59,000
Haemoglobin	Blood (horse)	68,000†
	Blood (man)	68,000
Hexokinase*	Yeast	97,000
Luciferase*	Firefly	100,000
Edestin	Hemp seed	320,000
Urease*	Soya bean	480,000
Adenosine triphosphatase*	Muscle	650,000
L-Glutamate dehydrogenase*	Liver	1,000,000
Haemocyanin	Edible snail	7,300,000 (approx.)

* Indicates that the protein possesses specific catalytic activity, i.e. the protein is an enzyme.

† Proteins for which the complete amino acid sequence is known.

34

positive and negative charges are equal; the net charge is zero but the actual charge is $\pm$. This corresponds to the isoelectric pH of a protein. In more acid solutions both amino acids and proteins carry net positive charges, i.e. more positive than negative charges, and in more alkaline solutions the number of negatives outweighs the number of positives so that the molecule as a whole is negatively charged.

The charges on the protein molecule arise from the R groups of certain of the constituent amino acids, some of which carry a second acidic group (e.g. glutamic acid) and others a second (e.g. lysine) or even a third basic group (e.g. arginine), over and above those that participate in the formation of peptide links. (Refer again to the formulae in Table 4 (pp. 27, 28).)

Thus a study of the simplest building units of which proteins are constructed provides us with a 'model' that enables us to understand some, at least, of the properties of proteins, in spite of the fact that while glycine and alanine have molecular weights of only 75 and 89 respectively, proteins are built up of many amino acid units and have molecular weights ranging from several thousands to a few millions. Table 5 gives some typical figures.

5

ENZYMES

Particularly important among the many proteins that go to make up the structure and mechanism of the living cell are the biological catalysts we call enzymes. These are very numerous. A recent book devoted to them lists about 1,000 such enzymes, and there is every reason to think that very many more will be added to the list as time goes on. About 150 have been isolated and purified sufficiently to show that they are proteins.

One of the earliest enzymes to be discovered, though its real nature could hardly be guessed at the time, was pepsin, a protein-splitting enzyme present in gastric juice. The great Italian physiologist, Spallanzani, fed hawks with pieces of meat enclosed in small wire boxes. When later the birds were persuaded to vomit, as is their wont, the boxes were found to be empty, showing that the birds' gastric juice contains some agent capable of liquefying meat. These experiments were done as long ago as 1783, but it was a long time before any more systematic work was done upon enzymes.

Three quarters of a century later, in 1858, the great Louis Pasteur, when he was studying the process of fermentation, showed that solutions of sugar are perfectly stable provided that they are sterile and air is excluded from them. But if air is allowed access to the solution, yeast cells from the air fall in and fermentation begins forthwith. Similarly, the souring of wine and milk were shown to be due to the activities of other kinds of micro-organisms, which Pasteur accordingly termed 'ferments' (1858). He believed that fermentation, souring and kindred processes involve and are inseparable from the life of the particular micro-organisms concerned.

Forty years later—as recently as 1897 to be precise—two German chemists, the brothers Buchner, were preparing a juice made by grinding up live yeast with sand and kieselguhr (a silicious earth) and squeezing the product in a hydraulic press. The resulting yeast juice was intended for medicinal use but it went bad very quickly and the problem of preserving it had to be dealt with. Among other things the Buchners tried the kitchen-chemistry method of adding large

Discovery and Nature of Enzymes

amounts of cane sugar. Without knowing it, they thus laid the foundations of the whole of our present-day knowledge of enzymes and enzymology, for their yeast juice, *although it contained no intact yeast cells whatsoever*, fermented sugar vigorously. Here, for the first time, an *enzyme* had been artificially separated from the cells which produced it.

This sugar-fermenting enzyme was given the name of 'enzyme', which literally means 'in yeast', but when later it was found that juices with other kinds of catalytic powers can be obtained by extraction from cells of other kinds, the name of 'enzyme' was taken over as a collective title and the yeast enzyme was given the distinguishing name of *zymase*.

Within a few years the newly discovered zymase was studied extensively, especially in England, by Harden and Young. They found among other things that *zymase loses its activity if boiled* and that it involves at least two component parts, an enzyme and a so-called coenzyme, the latter being stable to heat. This can be shown by the method of *dialysis*. If yeast juice is placed in a collodion or cellophane sac and the sac is placed in a large jar of distilled water, the small-molecular coenzyme passes through the cellophane into the water outside while the protein molecules are too large to pass through the pores of the membrane and are retained. The enzyme, freed in this way from small-molecular substances, is inactive, but activity returns if the water from the jar is concentrated and added back to the enzyme. Another way of restoring the activity is to add to the dialysed enzyme a little boiled yeast juice. Boiling destroys the enzyme but leaves the small-molecular coenzyme, *cozymase*, unaffected.

We know today that 'zymase' is not a single enzyme but a complex mixture of about 15 and that 'cozymase' is not a single substance but at least 4. The conversion of glucose or sucrose into alcohol and carbon dioxide by fermentation is not in fact a simple, one-step process but a long chain of stepwise chemical events, none of which proceeds at any appreciable speed except in the presence of the appropriate enzyme. Several of these require coenzymes which must also be present if the reactions are to proceed.

Even the early work of Harden and Young brought to light many of the principal and characteristic properties of enzymes, which may be defined as follows. *Enzymes, it may be said, are complex organic catalysts produced by living cells but capable of acting independently of the cells that produce them. They are characteristically thermolabile, i.e. destroyed by heat, and highly specific, i.e. their action is*

37

confined to one reaction or to a small group of similar chemical reactions. Many enzymes are inactive in the absence of a suitable *cofactor* or *coenzyme*.

PROPERTIES OF ENZYMES

Enzymes and coenzymes alike are true *catalysts*. The amounts in which they are required are minute in comparison with the amounts of chemical change that can take place in their presence. This is because, like catalysts of other kinds, they actually participate in the reactions they catalyse and are regenerated at the end of the process and so can be used over and over again.

Like other catalysts, enzymes can easily be 'poisoned' with the result that a part or the whole of their catalytic activity is lost, in which case we speak of their *inhibition* or *inactivation* respectively. Among the physical factors that have this effect are high temperatures, extremes of pH, ultraviolet light, violent mechanical agitation and many more. Inhibition or inactivation is frequently brought about by chemical reagents that react with $-SH$, $-NH_2$ or $-COOH$ groupings, suggesting that groups such as these play some vital part in enzymic catalysis, and indeed there is abundant evidence that this is the case. Other potent inhibitors include the salts of heavy metals, and acids that give rise to heavy ions in solution, such as phosphotungstic, perchloric and trichloroacetic acids. Heavy ions of the latter type are, of course, negatively charged and combine with enzymes in acid solutions, where enzyme proteins are positively charged, and form insoluble, salt-like complexes which are catalytically inert. Heavy metals exert their influence in solutions alkaline to the isoelectric pH of the enzyme protein, where the latter is negatively charged, and once again insoluble, catalytically inert, salt-like products are formed.

There are indications here that enzymes are indeed proteins, for nearly all the reagents that inactivate enzymes either precipitate proteins or produce in them subtle chemical alterations collectively known as 'denaturation'. It is not possible to say categorically that all enzymes are proteins because not all known enzymes have as yet been isolated. However, many have been crystallized and isolated in the pure state and in every case so far they have proved to consist of protein material.* When we add to this the close similarity between

---

* It should be noticed that crystallinity is no criterion of purity where proteins are concerned. They commonly form mixed crystals with contaminating proteins.

enzymes and proteins in their responses to inhibitory factors it does indeed seem more than probable that enzymes generally do indeed consist of proteinaceous material.

One important consequence of this is that, being proteins, enzymes have very high molecular weights and therefore their *molar* concentrations in cells and tissues are always very small. Think, for example of urease, an enzyme which catalyses the hydrolysis of urea:

$$CO\begin{array}{c} \diagup NH_2 \\ \\ \diagdown NH_2 \end{array} + H_2O \rightarrow CO_2 + 2NH_3.$$

Urea has a molecular weight of 60, whereas that of urease is about 480,000. If we were to prepare 1 % solutions of each, the *molar* concentration of the urea would be no less than 8,000 times as great as that of the enzyme. Many enzymes occur in cells and tissues in concentrations well on the way towards being infinitesimal and yet it is upon their catalytic activity that the very life of the cells depends.

One of the most remarkable features of enzymes, and one which distinguishes them from other and more familiar catalysts such as platinum black and other finely divided metals, is their very marked *specificity*. As is well known, platinum black will catalyse a fairly wide variety of different chemical processes, but a great many enzymes are *absolutely specific*. By this we mean that an enzyme of this kind can catalyse one reaction and one reaction only. In a smaller number of cases a given enzyme can catalyse any of a group of similar reactions, in which case it is said to have *group specificity*. The case of so-called *low specificity* is comparatively rare; one of the few known cases is found in some ester-splitting enzymes, where the enzyme, provided it is given an ester linkage to attack, pays little attention to the nature either of the alcohol or of the acid from which the ester is formed.

MEASUREMENT OF ENZYME ACTIVITY

If an enzyme is incubated together with its *substrate** under constant conditions of temperature and pH, the amount of chemical change taking place can be measured by some suitable analytical method. Suppose, for example, that the substrate is the non-reducing sugar, sucrose, and that the enzyme is yeast saccharase, a hydrolytic enzyme. The products of hydrolysis are glucose and fructose, both of

---

* The substance upon which its catalytic effect is exerted.

which have reducing properties, and the course of the reaction can therefore be followed by measuring the reducing power of samples withdrawn from the reaction mixture at known successive times. By plotting reducing power against time we can then arrive at what is known as the *progress curve* of the reaction. In most cases this curve has the form illustrated in Fig. 8.

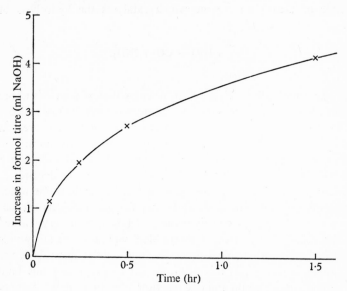

Fig. 8. Progress curve of an enzyme-catalysed reaction: digestion of casein by trypsin.

The reaction begins as soon as the enzyme is added to its substrate. For a time the curve is practically linear but the rate of reaction soon begins to fall off. It might have been expected that the rate would remain constant until the substrate is exhausted, but experience shows otherwise. As the reaction proceeds, many changes take place in the reaction mixture. Substrate is used up; the products appear and begin to accumulate and, if the reaction is a reversible one, will tend to oppose the forward reaction. In some reactions there may even be a change in pH. All of these changes influence the activity of the enzyme.

If, therefore, we wish to study the effects of such factors as temperature and pH upon enzyme activity, we must either see to it that the changes in the reaction mixture are the same in every experiment, or else take steps to minimize them. One way of doing this is to measure

Measurement of Enzyme Activity

the time taken by the enzyme to catalyse a given amount of chemical change. The time taken will then give a reciprocal measure of the enzyme's activity; reciprocal because an enzyme that is half as active will take twice as long to catalyse the same amount of chemical change. This method is satisfactory provided that the enzyme is stable under the experimental conditions but this requirement is not always satisfied and an alternative and preferable method is usually employed. Since changes in the reaction mixture start to take place very soon after the reaction begins, we must either make allowances for them, which is usually difficult, or else we must work over a very short time interval after the start of the reaction. If this interval is short enough the changes in the reaction mixture will be small enough to be negligible. Needless to say, if the interval is very short, the amount of substrate decomposed will also be very small, but many excellent micro-analytical methods are available today, so that accurate measurements of the initial reaction velocity can readily be made in most cases. The measurement of the initial reaction velocity of an enzyme-catalysed reaction gives a reliable measurement of the enzyme's activity under any given set of experimental conditions.

INFLUENCE OF pH

The influence of pH upon enzyme activity is very marked, and the vast majority of enzymes are active over only a very narrow range of pH. Fig. 9 shows some typical activity curves. Under a given set of conditions there is a well marked peak of activity at what is called the *optimum pH*, and the activity falls off sharply with changes of pH on either side of the optimum value. The optimum pH is not a fixed and unalterable property but can vary with such things as the ionic strength and nature of the buffer used and by the length of time for which the enzyme is allowed to act.

INFLUENCE OF TEMPERATURE

Temperature too has a powerful effect upon enzymes. It is a general rule that chemical reactions go faster at higher temperatures, and enzyme-catalysed reactions are no exception. However, in the case of enzyme-catalysed processes, not only does the reaction go faster at higher temperatures but the thermal destruction of the enzyme also takes place more rapidly. Consequently for any given period of time there will be an *optimum temperature*, i.e. a temperature at which the greatest amount of chemical change is brought about in a given time by a given amount of enzyme under a given set of experimental conditions.

Enzymes

REVERSIBILITY OF ENZYME ACTION

Theoretically all chemical reactions are reversible and a catalyst that accelerates a reaction in one direction ought, again theoretically, to accelerate it in the other direction also, so that all enzymically catalysed reactions should tend towards equilibrium. This reversibility of enzyme activity has indeed been demonstrated in many cases, and Fig. 10 shows the results of experiments on the action of a fat-

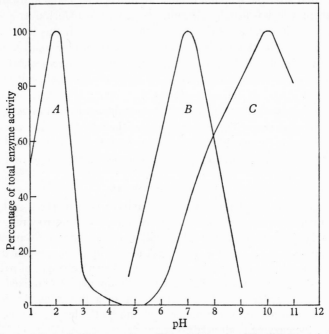

Fig. 9. Effect of pH on enzyme activity. (*A*) Pepsin; (*B*) salivary amylase; (*C*) arginase. (Modified after Fruton and Simmonds.)

splitting enzyme (castor oil bean lipase) on the hydrolysis and synthesis of a typical fat: the final equilibrium mixture is the same whether we start with the ester or with its separate components. In many cases, however, the equilibrium is so far over towards one side or the other that, for all practical purposes, the reaction is unidirectional. An excellent example of this is the decomposition of hydrogen peroxide which is catalysed by an iron-containing enzyme called *catalase*;

$$2H_2O_2 \rightarrow 2H_2O + O_2$$

CLASSIFICATION OF ENZYMES

The classification of enzymes is not an easy matter if only because they catalyse between them such an immense range and variety of chemical reactions. To make matters worse, no fool-proof method for naming these catalysts was invented until very recently. In the early days it was the practice to add -*ase* to the name of the substrate, i.e. the substance upon which the enzyme exerts its activity, but this

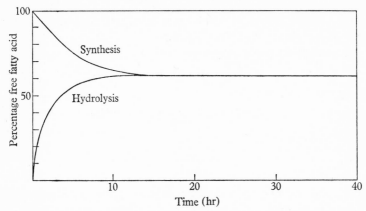

Fig. 10. Reversibility of action of lipase of *Ricinus* on triolein.

no longer suffices. Ideally the name given to an enzyme should indicate several things; first the *identity of the substrate*, secondly *the kind of chemical change* wrought upon that substrate and thirdly an indication of *the origin of the enzyme*. Thus we might speak of 'muscle lactate dehydrogenase'. This would signify an enzyme which (*a*) acts upon lactate, (*b*) catalyses the removal of hydrogen from it, and (*c*) is present in muscle. Unfortunately, it is only in a few cases that the names in current use give so much information.

Since classification seems to be necessary, and since an internationally agreed method of classification has recently been established, we can now use the currently accepted system and give a few simple examples. The list of examples can be extended by the reader himself as he goes through the pages that remain of this book.

The classification proposed runs briefly as follows:

(*a*) HYDROLASES. Enzymes which catalyse a splitting of their substrates with the aid of a second reactant, usually water, so that the reaction catalysed is a hydrolytic one.

Examples—urease:

$$CO\begin{cases}NH_2\\[2pt]\\[2pt]NH_2\end{cases} + H_2O \rightarrow CO_2 + 2NH_3;$$

amylase:

$$starch + H_2O \rightarrow maltose;$$

maltase:

$$maltose + H_2O \rightarrow 2\ glucose.$$

(*b*) LYASES. Enzymes which resemble (*a*) but require no second reactant to act as a 'splitting' agent.

Examples—carbonic anhydrase:

$$H_2CO_3 \rightleftharpoons H_2O + CO_2;$$

carboxylase:

$$CH_3CO.COOH \rightarrow CH_3CHO + CO_2;$$
$$(pyruvic\ acid)$$

catalase:

$$2H_2O_2 \rightarrow 2H_2O + O_2.$$

(*c*) TRANSFERASES. These catalyse the transference of a group of some kind (R) from a donator molecule (A.R.) to an acceptor molecule (B) thus:

$$A.R + B \rightleftharpoons B.R + A.$$

Examples will be found in later sections (e.g. *transaminase*, p. 48).

(*d*) ISOMERASES; catalysing intramolecular rearrangements in the substrate molecule.

Example—triose phosphate isomerase

$$\begin{array}{ccc}
CH_2O.PO_3H_2 & & CH_2O.PO_3H_2 \\
| & & | \\
CHOH & \rightleftharpoons & CO \\
| & & | \\
CHO & & CH_2OH \\
glyceraldehyde & & dihydroxyacetone \\
phosphate & & phosphate
\end{array}$$

(*e*) OXIDOREDUCTASES. This group includes several groups of oxidizing and reducing enzymes.

Example—lactate dehydrogenase.

$$CH_3CH(OH)COOH \rightleftharpoons 2H + CH_3CO.COOH.$$

The 2H are handed over to an appropriate 'hydrogen acceptor'.

(*f*) LIGASES or SYNTHETASES. Enzymes which catalyse the synthesis of many biological substances.

FUNCTIONS OF ENZYMES

The first chemical event that happens to food taken by an animal is that it is *digested*. This process involves a fairly large number of hydrolases, each specific for the hydrolysis of its own kind of substrate. *All the digestive enzymes are hydrolases,* i.e. they catalyse hydrolytic processes. There are, for example, *amylases* (from Latin, *amylum* = starch) which catalyse the hydrolytic breakdown of starch, and *lipases*, which act similarly upon fats. The protein-splitting enzymes are called *peptidases* and the net result of digestion is that the complex molecules that go to make up the food are resolved into their simpler component parts, monosaccharides, fatty acids and amino acids. These are absorbed from the intestines and enter the blood stream.

By way of the blood these simpler materials are distributed to the various organs of the body and enter a vast and exceedingly intricate maze of enzymically-catalysed chemical pathways to which the collective name of *metabolism* is given. Some of these reactions are synthetic in nature and others are reactions of degradation to form simpler substances, the eventual end-products of which are mainly water, carbon dioxide and, in mammals, urea. Metabolism as a whole can therefore be subdivided into *anabolism*, which covers reactions leading to synthesis, e.g. of body proteins from amino acids, and *katabolism*, which includes reactions that lead to degradation; of sugar to CO_2 and water for instance.

It is easy enough to break down sugar or butter into carbon dioxide and water by simply throwing them on the fire. But this method is very wasteful of energy, for the energy set free in this form of combustion will merely add a little to the heat emitted by the fire. Most of it will go up the chimney in any case. In a living organism things are very different. The organism lives on energy derived from the katabolism of its food, and the katabolic processes through which the organism gets access to this energy are devious, step-by-step operations very unlike those going on in a fire. At each and every one of the intermediate stages an enzyme, and where necessary a coenzyme also, is required. None of these processes is wholly independent of the rest, but all are interlinked with other lines of metabolism so that, in effect, the total metabolism of an organism is a very complex affair, full of interlinking processes. To unravel this network as completely as possible is among the chief aims and

Enzymes

ambitions of the biochemist. Much has already been done in this direction and some of the results will appear in the chapters that follow.

DIGESTION OF PROTEINS

The principal secretions involved in digestion are: (*a*) *saliva*, secreted by the salivary glands; (*b*) *gastric juice*, which is secreted by the cells lining the stomach and contains a good deal of free hydrochloric acid; (*c*) the *pancreatic juice* and (*d*) the *bile* secreted by the pancreas and liver respectively; and (*e*) an *intestinal juice* elaborated by the mucous lining of the small intestine.

The first hydrolase to attack the food protein is *pepsin*, a proteolytic enzyme contained in the acid gastric juice. This enzyme has an unusually acid optimum pH of about 1·5–2, which is provided by the HCl of the juice. Pepsin provides a good example of a phenomenon we have not so far mentioned. As secreted it has no action upon proteins because the active parts of the enzyme are masked, so to speak, by some other substance. This 'mask' is removed by the free HCl of the gastric juice and the activity of the enzyme is released. Once some free pepsin has been produced, it unmasks and so activates more of its precursor (*pepsinogen*) so that, once begun, activation is an autocatalytic process.

Pepsin is classified as an *endopeptidase*, that is to say, it attacks the protein chain at certain points remote from the ends of the peptide chain. It is highly specific and attacks only linkages between certain particular amino acids, thus breaking up the long chain into shorter fragments.

When the partly digested mass passes on into the duodenum it encounters the pancreatic juice and bile which between them contain about enough free alkali to neutralize the acid coming through from the stomach. The pancreatic juice contains another enzyme precursor, *trypsinogen*, which is activated by a different kind of 'unmasking' and yields free *trypsin*, another endopeptidase. This enzyme carries on the work begun by pepsin, but differs from pepsin in specificity and attacks at different and specific linkages, producing yet shorter fragments of amino acid chains.

The final stages of attack are carried out by a group of hydrolytic enzymes present in the intestinal juice to which the collective name of *erepsin* is usually given. At least three groups of enzymes are present here and all are *exopeptidases*, i.e. they attack at the ends of the

46

fragmented chains. One group attacks at the end where there is a free carboxyl group, and a second at the end carrying a free amino group. This process of chipping away terminal units continues until only two amino acids remain in each fragment. These residual dipeptides are immune against attack by the other enzymes of the erepsin complex and a third group of enzymes, the *dipeptidases*, complete the work by hydrolysing the dipeptides.

In the end, therefore, all the amino acids which were linked together in the original food protein are set free, undergo absorption through the walls of the small intestine, and enter the blood stream. They are distributed to the body generally, each tissue apparently taking up what it needs at the time. Any surplus amino acids go on to lose their amino groups, the nitrogen they contain being converted into urea, which is subsequently excreted in the urine. The residues of these surplus amino acids, once their nitrogen has been removed, contribute to the stores of carbohydrate and fats.

At the end of the processes of digestion and absorption and the initial stages of metabolism, then, the tissues have been provided with amino acids to meet their immediate needs, surplus amino acids have been deaminated, and the deaminated residues have been put away on one side to await metabolism for energy production when the time comes.

6

FATE OF AMINO ACID NITROGEN

TRANSDEAMINATION

Surplus amino acids, i.e. amino acids over and above those required to fulfil the animal's immediate needs, undergo a process known as *deamination*, in other words, they lose their amino groups. Later, the nitrogen contained in these groups appears in the urine in the form of urea. Urea is characteristic of mammals, but in birds and most reptiles its place is taken by uric acid, a point to which we shall return presently. Let us for the moment consider the manner in which the $-NH_2$ groups are removed from the amino acids and how they are converted into urea.

We know that the liver plays a large part in urea formation because if the liver is removed from a dog or if the blood supply of the liver is short circuited, the animal survives for a few days, but during that time it makes no more urea. Moreover, it is possible to remove the liver from a dog, fix it up with an artificial circulation, and add amino acids to the circulating blood and then we find that urea is produced at the expense of the amino acids. Clearly, then, it is in the liver that urea is formed. Further experiments show that urea can also be made from ammonia, again by the perfused liver, which suggests that the processes we have to consider are these:

amino acids → ammonia → urea.

After many years of research the mechanism of deamination was brought to light by Braunstein and Kritzmann. It involves the action of a transferring enzyme and the participation of an α-keto acid. The general reaction is of the following type:

$$R_1.CH(NH_2)COOH + R_2.CO.COOH \rightarrow$$
$$R_1.CO.COOH + R_2.CH(NH_2)COOH.$$

The nature of this transfer reaction is more clearly apparent if it is written in the following manner:

$$
\begin{array}{cc}
R_1 & R_2 \\
| & | \\
CH(NH_2) & CO \\
| & | \\
COOH & COOH \\
& \\
R_1 & R_2 \\
| & | \\
CO & CH(NH_2) \\
| & | \\
COOH & COOH \\
\end{array}
$$

Transdeamination

The enzyme concerned is called a *transaminase*, for reasons that are sufficiently obvious, and the amino group acceptor, represented in the equation by $R_2.CO.COOH$, is a keto acid familiar enough to biochemists under the name of α-ketoglutaric (α-oxoglutaric) acid:

$$
\begin{array}{ll}
\text{COOH} & \text{COOH} \\
| & | \\
\text{CH}_2 & \text{CH}_2 \\
| & | \\
\text{CH}_2 & \text{CH}_2 \\
| & | \\
\text{CO} & \text{CH(NH}_2) \\
| & | \\
\text{COOH} & \text{COOH} \\
\textit{α-ketoglutaric acid} & \text{glutamic acid} \\
& \textit{(α-aminoglutaric acid)}
\end{array}
$$

The products of this transfer are, first, the α-keto acid corresponding to the amino acid with which we started and, secondly, glutamic acid. This latter is one of the amino acids which enter into the structure of proteins and which, by the way, is a member of the non-essential group. The newly formed α-keto acid goes off on its way to be converted into fuel for the metabolic furnaces, and need not concern us any more at the moment. The fate of the glutamic acid is of more interest at this stage of our inquiries.

The liver contains an oxidoreductase, *glutamate dehydrogenase*, which can split off the $-NH_2$ from this amino acid, yielding ammonia and regenerating the α-ketoglutaric acid with which we started. This enzyme requires a coenzyme and the reaction can be written thus:

$$
\begin{array}{ll}
\text{COOH} & \text{COOH} \\
| & | \\
\text{CH}_2 & \text{CH}_2 \\
| & | \\
\text{CH}_2 & \text{CH}_2 \\
| & | \\
\text{CH(NH}_2) + \text{coenzyme} \rightarrow & \text{C}{=}\text{NH} + \text{coenzyme}.\text{H}_2 \\
| & | \\
\text{COOH} & \text{COOH} \\
\textit{glutamic acid} & \textit{α-iminoglutaric acid} \\
\textit{(α-aminoglutaric acid)} &
\end{array}
$$

This first stage of the process of deamination is followed by one of the few metabolic reactions we know that appear to require no enzyme, for the α-imino acid formed is exceedingly unstable and reacts, apparently spontaneously, with water, to give free ammonia and α-ketoglutaric acid;

$$
\begin{array}{ll}
\text{COOH} & \text{COOH} \\
| & | \\
\text{CH}_2 & \text{CH}_2 \\
| & | \\
\text{CH}_2 & \text{CH}_2 \\
| & | \\
\text{C}{=}\text{NH} + \text{H}_2\text{O} \rightarrow & \text{CO} + \text{NH}_3 \\
| & | \\
\text{COOH} & \text{COOH} \\
\textit{α-iminoglutaric acid} & \textit{α-ketoglutaric acid}
\end{array}
$$

49

This somewhat complicated mechanism calls for several comments. It is essentially an oxidative process in which 2 atoms of hydrogen are removed from the starting material and handed over to the coenzyme. From there they are passed on along a chain of so-called hydrogen carriers and react eventually with oxygen to form water. This is a fairly typical example of the way in which organic substances are oxidized in living stuff and we shall deal with reactions of this sort

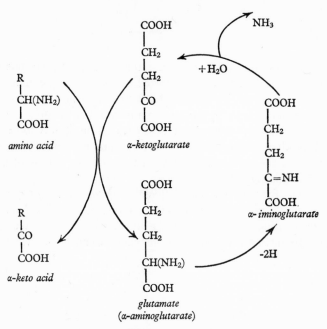

Fig. 11. Mechanisms involved in transdeamination of amino acids. See text for detailed explanation.

in more detail in a later chapter. The second interesting point can best be appreciated if we combine the foregoing equations in the form of the diagram shown in Fig. 11. It will be noticed that a single molecule of α-ketoglutaric acid can be used over and over again, deaminating one molecule of amino acid at each turn of the cycle, producing one molecule of ammonia and being regenerated for use over again. It is, in fact, a catalyst, as truly a catalyst as the enzymes which speed up the whole process. So much for the problem of transdeamination; we can turn next to that of urea formation.

FORMATION OF UREA

As the experiments with the dog and its liver show, urea formation takes place only in the liver. Remove the liver and urea formation ceases completely. The elucidation of this problem was achieved by Krebs, followed by many others who worked out the finer details of the process, though these need not concern us here. Krebs worked with very thin slices of liver tissue, slices no thicker than about 0·3 mm. If slices as thin as this are suspended in warm, well oxygenated Ringer's solution, they survive for several hours and reproduce rather faithfully the activities of the liver itself.

Krebs showed first of all that if such slices are provided with small amounts of glucose to serve as 'fuel' they will produce urea from added ammonia and bicarbonate. Next he tried various amino acids as sources for the ammonia, since these were already known to be deaminated by liver tissue. Entirely unexpected results followed when he tried the amino acid *ornithine* (formula below) in the presence of small amounts of ammonia. With this amino acid the rate of urea synthesis increased by about ten times; ornithine evidently exerts a catalytic effect in some way. Now it had long been known that the liver contains *arginase*, an enzyme which catalyses the hydrolysis of another amino acid, arginine, to give ornithine and urea:

$$
\begin{array}{l}
\text{HN=C}\!\!\begin{array}{l}\nearrow \text{NH}_2 \\ \searrow \text{NH}\end{array} \\
\quad\quad | \\
\quad\ (\text{CH}_2)_3 \\
\quad\quad | \\
\quad\ \text{CH(NH}_2) \\
\quad\quad | \\
\quad\ \text{COOH} \\
\quad\ \textit{arginine}
\end{array}
\;+ \text{H}_2\text{O} \longrightarrow\;
\begin{array}{l}
\text{NH}_2 \\
| \\
(\text{CH}_2)_3 \\
| \\
\text{CH(NH}_2) \\
| \\
\text{COOH} \\
\textit{ornithine}
\end{array}
\;+\;\begin{array}{l}\text{CO(NH}_2)_2 \\[2.2em] \textit{urea}\end{array}
$$

Krebs argued that, if ornithine could in some way add on ammonia and carbon dioxide and so be converted into arginine, arginase would split the product, giving urea and regenerating ornithine to be used all over again. This would provide an explanation for the undoubtedly catalytic action of ornithine.

The drawback about this pretty idea was that, as a glance at the formulae will show, it is rather a far cry from ornithine to arginine. Was any substance known that could act perhaps as a half-way house between the two compounds? Ever resourceful, Krebs turned

51

to another amino acid, *citrulline*, so-called because it had been discovered some years earlier in *Citrullus vulgaris*, the water melon. The structure of citrulline lies midway between that of ornithine and that of arginine:

$$
\begin{array}{c}
NH_2 \\
\diagup \\
O{=}C \\
\diagdown \\
NH \\
| \\
(CH_2)_3 \\
| \\
CH(NH_2) \\
| \\
COOH
\end{array}
$$

citrulline

Repeating his experiments in the presence of citrulline in place of ornithine he found that this compound too has a catalytic action on urea synthesis. Taking all these results together it was now possible to produce a cyclical scheme, the 'ornithine cycle' as it came to be known, as shown in Fig. 12. This has found almost universal acceptance. Ornithine takes on ammonia and carbon dioxide to yield citrulline, which takes on more ammonia and yields arginine. Arginine is then split by arginase, giving urea and regenerating ornithine.

Here is another case in which a small amount of a substance can be used over and over again, just as we saw in deamination. Many similar examples are known today and it is customary to refer to substances which behave like ornithine and α-ketoglutaric acid (p. 50) as 'carrier catalysts'.

By way of rounding off this story it may be said that the ornithine cycle has proved on further study to be a good deal more complex than appears on the surface, and most of the intermediate steps between ornithine and citrulline, and between the latter and arginine, have been worked out in detail but these details are too numerous and too complex to be discussed here. They can be found, however, in any standard book on biochemistry.

COMPARATIVE ASPECTS OF NITROGEN EXCRETION

It was mentioned earlier that while urea is the characteristic nitrogenous end-product of protein metabolism in mammals, it is replaced in birds and reptiles by *uric acid*. Much work has gone into efforts to find reasons for these differences. Most aquatic animals, including many though not all fishes, merely excrete ammonia without building it up into more complex products like urea and uric acid. But as soon as we reach the amphibia, which spend a part of their time on the land, urea appears as a major nitrogenous end-product. The mammals, which had an amphibian ancestry, have retained urea as their end-product, but somewhere among the reptiles, which gave rise to the birds in the course of evolution, a change took place leading to the replacement of urea by uric acid.

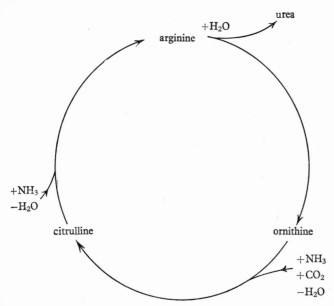

Fig. 12. Synthesis of urea by the 'ornithine cycle'. See text for explanation.

But let us take first things first. Why do some animals synthesize one and some the other of these two end-products? The mechanisms involved in their synthesis are very complex and use up a good deal of energy, and it is unlikely that they would have originated in the course of evolution and been retained thereafter without some good reason.

The answer appears to be that ammonia is a very poisonous substance, and that so long as animals lived entirely in the water they could dispose of it as fast as it is produced. But animals could spend a part or the whole of their life on the land, only with the advent of chance mutations, which provided them with some means of detoxicating ammonia formed by deamination of incoming amino acids; otherwise their lives would have been in jeopardy every time they ate a protein meal.

Fate of Amino Acid Nitrogen

Ammonia is so familiar a household article that its toxicity is not usually realized. But that it is very poisonous there can be no doubt. This has been very ingeniously demonstrated in rabbits by Sumner, who was the first to crystallize an enzyme in the pure state. The enzyme in question was urease, which catalyses the hydrolysis of urea to give carbon dioxide and ammonia. Sumner injected purified urease into rabbits, the blood of which contains small amounts of urea. This urea was hydrolysed by the enzyme, yielding carbon dioxide and ammonia, and the animals died as soon as the ammonia content of the blood had risen to about 1 part in 20,000—a very high order of toxicity indeed. It could of course be objected that perhaps urease itself is toxic, but Sumner ruled out this possibility by injecting his enzyme into birds, the blood of which contains no urea. The birds were unharmed and went on their way. However, when urease was injected together with urea, the birds died when the ammonia level in their blood was still at a very low level. Of the toxicity of ammonia there can be no doubt.

The change over from the excretion of ammonia to that of urea was a very important step in the evolution of the amphibians, for no animal could spend much time on dry land without being exposed to the dangers of ammonia poisoning. But by converting waste ammonia into urea this danger was overcome.

Although most reptiles and all birds subsequently abandoned the habit of urea formation which they no doubt inherited from amphibious ancestors, urea formation is still carried on by the mammals, including even the few primitive mammals that still lay eggs. Typically, however, the modern mammalian embryo develops in the amnion, which is virtually a private pond, and has all the maternal water supply at its disposal. Urea formed by the embryo can pass across the placenta and into the blood stream of the mother, to be excreted by proxy by the maternal kidneys. Urea still serves as a nitrogenous waste product for mammalian embryos and adults alike.

Now why, with the evolution of the reptiles and birds did the synthesis of uric acid replace that of urea? The answer proposed by Needham is that, with the evolution of the shelled egg, which is characteristic of reptiles and birds, ammonia production in the course of development would be unthinkable and the production of urea would also be undesirable because it would bind up a considerable part of the very limited supply of water with which eggs of this kind are provided when they are laid. Shelled eggs could not therefore be expected to work until some other way of disposing of ammonia had been developed. Uric acid seems to have provided the answer; it is relatively innocuous, like urea; it is too weak an acid to upset the pH of the shell contents; and, being very insoluble indeed, would not interfere with the water relationships of the developing embryo. Once discovered and installed, the mechanisms for uric acid synthesis were carried on from embryonic to adult life among the reptiles and handed on to the birds.

Some work has been done to find out at what point in evolution the changeover took place, and with interesting results. All the birds, snakes and lizards that have been examined produce uric acid, with no significant amounts of urea. Crocodiles and alligators are interesting because they spend most of their lives in water and, while they produce some urea and a good deal of uric acid, their urine also contains a high proportion of ammonia. This is probably a satisfactory arrangement so long as they can spend most of their time in water. We do not know how they get on when the water dries up in the dry season and they have to undertake overland migrations in search of fresh haunts. But we know a good deal more about tortoises and turtles. All these reptiles, at any rate as far as they have been studied, make urea with a few significant exceptions. This group probably had an amphibious origin, but some have returned to water and, though they do so on a smaller scale than those which remained amphibious, they still make significant amounts of urea as evidence of their evolutionary origin. Other members of this ancient group have moved away from amphibious conditions in favour of drier environments and among these the most interesting facts have been discovered. In Table 6 are shown the amounts of nitrogen excreted as ammonia, urea and uric acid, expressed as percentages of the total amount of nitrogen excreted. The data show that the wholly aquatic species produce approximately equal amounts of urea and ammonia. Amphibious species excrete much urea and only a little ammonia; a dry-living species produces uric acid, together still with a significant quantity of urea, while a desert-living form has practically abandoned urea formation in favour of that of uric acid.

Presumably then, in the early days of evolution, reptiles still lived an amphibious kind of life and carried on the amphibian tradition of urea formation. Later on, apparently, they took to drier habitats and discovered the advantage of uric acid as an end-product, and experimented with it for some time before discarding the old and well tried device of urea formation. Presently with the development of efficient shelled eggs they dropped urea production once and for all and became the whole-hearted producers of uric acid that they, and their descendants, the birds, are today. It is not often that we can find a metabolic change-over still in evidence as we so fortunately have in the tortoises and turtles.

Table 6. *Nitrogen excretion of some chelonian reptiles*
(Rough averages)

Habitat	Percentage of total non-protein N as		
	Ammonia	Urea	Uric acid
Aquatic	20–25	20–25	5
Amphibious	6	50–60	5
Terrestrial, dry living	6	20–30	50
Terrestrial, desert living	5	5–10	50–60

The change over from urea excretion to that of uric acid entailed major changes in the structure of the excretory apparatus of animals. The kidneys of birds and the majority of reptiles are specially adapted for the excretion of a waste product as insoluble as uric acid while the mammalian kidney is not. The large-scale formation of uric acid is, fortunately, not necessary in the mammals and, indeed, when it is formed in unusually large amounts as a result of dietary indiscretions it only leads to gout and misery.

7

CARBOHYDRATES

The carbohydrates with which we have to deal here can be broadly divided into simple sugars and high polymers of simple sugars, such as starch and cellulose. *Cellulose* and certain other high polymers play parts of the greatest structural importance in plants and even in a few animals, for in the tunicates (sea-squirts) the body wall is made of cellulose. In general, however, cellulose is of little importance among animals except as a food for herbivores. *Starch*, on the other hand, plays little or no part in structure either in animals or plants, but is of the highest importance as a store of carbohydrate material in plants of many kinds, especially in Monocotyledons, and as a food source of carbohydrate for animals. In the Compositae, e.g. *Dahlia* and the Jerusalem artichoke, starch, which is a polymer of glucose, is replaced by a polymerized fructose, *inulin*, which again serves as a store of carbohydrate. Grasses contain other polymers of fructose known as *levans*.

Animals produce a glucose polymer known as *glycogen*, which acts as a store and resembles starch in structure, though the molecules are considerably smaller than those of starch. This substance is stored mainly in the liver and muscles. Other tissues, with a probable exception in the case of brain tissue, also contain reserves of carbohydrate in the form of glycogen; and the glycogen of the peripheral tissues is built up from blood glucose, which is formed in its turn by the breakdown of the glycogen of the central stores contained in the liver.

Much work has been done on the structures of these polysaccharides and a good deal is known about them. Starch, for example, has been found to consist of two components, the first of which, *amylose*, consists of a long chain of several hundreds of glucose units strung together end to end. It is this substance that gives the familiar deep, pure blue coloration with iodine. In the other component, *amylopectin*, the number of glucose units is even larger, though the individual chains are shorter and are very highly branched, and give

Carbohydrates

only a somewhat dirty violet colour with iodine. A diagrammatic representation is shown in Fig. 13. Animal glycogen resembles this second substance, amylopectin, but has no counterpart of amylose.

The simpler sugars are of great importance to animals and plants alike. There are two main types of simple monosaccharide sugars, the

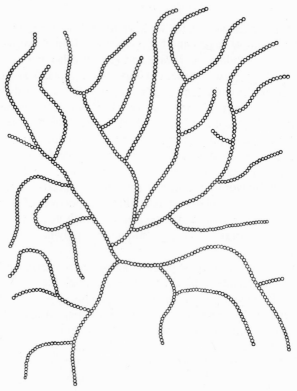

Fig. 13. Diagram to illustrate general structure of amylopectin and glycogen. Each circle represents a glucose residue.

aldoses, which contain an aldehydic reducing group, and the ketoses, which contain a ketonic group. *Glucose*, with which the reader will probably be familiar enough, is an example of an aldohexose, while *fructose* is a well-known ketohexose. These are 6-carbon sugars, but other sugars with from three to seven carbon atoms are all important in plant and animal metabolism. For example, the 7-carbon sugar, *sedoheptulose*, plays an important part in photosynthesis. It may well be that even larger monosaccharide molecules play some part in metabolism but, if so, little is at present known about them.

MONOSACCHARIDES; STRUCTURE AND PROPERTIES

Let us consider the simplest of the common sugars for a moment, namely, the two trioses, *glyceraldehyde* and *dihydroxyacetone*.

$$
\begin{array}{cc}
\text{CHO} & \text{CH}_2\text{OH} \\
| & | \\
\text{HC.OH} & \text{CO} \\
| & | \\
\text{CH}_2\text{OH} & \text{CH}_2\text{OH} \\
\textit{glyceraldehyde} & \textit{dihydroxyacetone}
\end{array}
$$

Alternative names for these would be aldotriose and ketotriose. Both have reducing properties. Now glyceraldehyde, though not the corresponding keto-compound, contains an asymmetric carbon atom so that two forms exist, a D- and an L-isomer*

$$
\begin{array}{cc}
\text{CHO} & \text{CHO} \\
| & | \\
\text{HC.OH} & \text{HO.CH} \\
| & | \\
\text{CH}_2\text{OH} & \text{CH}_2\text{OH} \\
\text{D-}\textit{isomer} & \text{L-}\textit{isomer}
\end{array}
$$

Higher aldose sugars, tetroses, pentoses, hexoses and heptoses can be considered as being derived from the trioses by inserting additional $-$CHOH units between the reducing and terminal alcoholic groupings, thus from D-glyceraldehyde we could get a series of sugars with the following general formula:

$$
\begin{array}{c}
\text{CHO} \\
| \\
\text{CHOH} \\
| \\
\text{CHOH} \\
| \\
\text{CHOH} \\
| \\
\text{*HC.OH} \\
| \\
\text{CH}_2\text{OH}
\end{array}
$$

* The prefixes D- and L- have no reference to the effect of the substances concerned upon polarized light, though it does happen that D-aldotriose is dextrorotatory and the L-isomer is laevorotatory. D- and L- are used only to indicate the spatial arrangement of the secondary CHOH group immediately adjacent to the terminal $-$CH$_2$OH.

Carbohydrates

All the natural aldose sugars can be formulated in this way, but in every one of them the positions of the H and OH at the C atom marked with the asterisk are always the same with reference to the terminal alcoholic group. All, therefore, are members of the D-family of sugars. Members of the L-family are occasionally met with, but they are as rare as the D-sugars are common.

In the hypothetical hexoses represented in the formula above there are in all four asymmetric carbon atoms of which only one is determinate, so that the total number of possible isomeric D-aldohexoses alone is rather large, but fortunately only two need concern us much and a third turns up only occasionally. These three are glucose, galactose, and mannose:

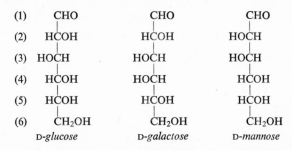

| | D-glucose | D-galactose | D-mannose |

Now, none of these formulae accounts completely for all the known properties of the sugars, and to illustrate this we may take the case of the commonest of sugars, D-glucose, even though this means straying for a time from the paths of biochemistry and going back to a slightly purer kind of chemistry.

If glucose is crystallized from boiling water the product melts at 156 °C and has a specific rotation of $[\alpha]_D = +19°$. Crystallization from glacial acetic acid yields a glucose with quite different physical properties, m.p. 146 °C, $[\alpha]_D = +110°$. If either of these forms of glucose is dissolved in water and the optical rotation is observed, it is found to change gradually from its starting value to a final $[\alpha]_D$ of $+52°$. This phenomenon is called *mutarotation* and is a property common to all the aldoses. Clearly, two different forms of glucose exist, and this can only be accounted for if there is one more asymmetric carbon atom than the straight-chain formula suggests. Much work on this problem has led to a revised formula which can account for the known facts; it involves the formation of a ring structure as shown at the top of the next page.

```
(1)    CHOH
        |        \
(2)    HCOH        \
        |            \
(3)    HOCH          O
        |            /
(4)    HCOH        /
        |        /
(5)    HC      /
        |
(6)    CH₂OH
```
D-*glucose*

The H and OH at carbon 1 can be arranged in either of two ways, one corresponding to what is called the α-sugar (that with the high$[\alpha]_D$), and the other to the β-form, (with the low $[\alpha]_D$);

```
H      OH              HO      H
  \   /                  \   /
   C                       C
  /\                      /\
 α-form                  β-form
```

The H and OH can change positions, probably through an opening and re-closing of the ring, and, if we start with either isomer alone in solution, exchange takes place and continues until an equilibrium is reached between the two forms: hence mutarotation.

The revised formula accounts at the same time for the fact that, although these sugars have reducing properties, they reduce much more feebly than they should if a free aldehydic group is present. Other substances with the same type of structure as is now accepted for glucose have a reducing power of the same order as that of glucose and the other aldohexoses.

One more piece of evidence needs to be taken into account and comes from X-ray crystallography. This shows that the *carbon atoms are not arranged spatially in a line but in a ring*; four carbon atoms and the oxygen of the ring lie practically in one plane with the appendages projecting above and below this plane. Today, therefore, the formula of glucose is usually written in the following manner:

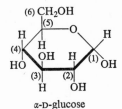

α-D-glucose

The numbering corresponds to that used in the earlier formulae.

61

Carbohydrates

One of the most important biological properties of the sugars is a condensation reaction that can take place between the −OH at carbon 1 with alcoholic substances. With methyl alcohol, for example, we can produce two methyl glucosides, one corresponding to α- and the other to β-glucose. One feature of these products is that, unlike the free sugars, they do not mutarotate, nor do they react with phenylhydrazine to form osazones; substitution of this particular hydroxyl group, the *glycosidic hydroxyl*, prevents the free interchange that leads to mutarotation in the parent sugar, and prevents osazone formation at the same time, presumably because it prevents opening and re-closing of the ring.

DISACCHARIDES

Glycoside formation is an interesting process because reactions of this kind can lead to the formation of disaccharides. Suppose that we have two molecules of glucose, one of which reacts with the other through its glycosidic hydroxyl group. The second molecule has hydroxyl groups at carbons 1, 2, 3, 4 and 6, any of which might enter the reaction. Moreover, the glycosidic hydroxyl of the first molecule might be in either the α- or the β-position, so that no less than ten different disaccharides can be formed from two molecules of D-glucose. In nature, however, only reactions involving the hydroxyls at carbons 1, 4 and 6 are at all common, and of these we will take only that involving carbon 4, since glycosides of this type are perhaps the commonest of all.

Two such sugars are possible:

α-D-*maltose* (α-glucosido-4-glucose)

α-D-*cellobiose* (β-glucosido-4-glucose)

62

Disaccharides

Like their parent sugars these disaccharides have an unsubstituted glycosidic hydroxyl group and accordingly have *reducing properties* and form *osazones*; moreover, and for the same reason, both *mutarotate* and both can form *glycosides* known as maltosides and cellobiosides respectively. This can lead on, of course, to the formation of tri- and higher saccharides. Both maltose and cellobiose exist in nature.

Similar general arguments arise in the case of fructose and again lead to a ring formula, but the situation is complicated a little by the fact that fructose, in biological combination, seems invariably to occur in a five- instead of the six-membered form which predominates in free fructose:

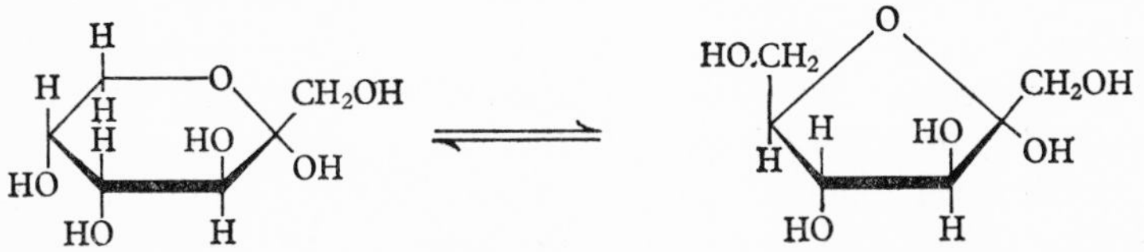

| (pyranose form) | α-D-*fructose* | (furanose form) |

(The names pyranose and furanose are taken from reference compounds containing these two types of ring structure.)

Like glucose, fructose forms glycosides (fructosides) and among these the most interesting is *sucrose* (cane sugar). This is formed by the union of a β-fructoside residue with an α-glucoside grouping. The union involves the glycosidic hydroxyls of both sugars so that the product has no free glycosidic group and does not reduce, form an osazone, or mutarotate;

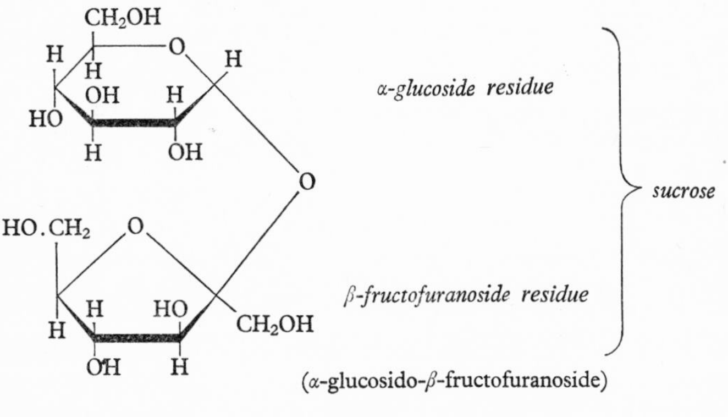

α-glucoside residue

β-fructofuranoside residue

(α-glucosido-β-fructofuranoside)

sucrose

63

Carbohydrates

It will be noticed that the fructose is present in its five-membered or furanose form. This five-membered form is very unstable and, in a solution of free fructose, reverts very rapidly to the more stable 6-membered pyranose isomer. Compounds which contain fructose in the furanose form inherit some of the instability of this sugar; for example, they are very easily hydrolyzed by dilute mineral acids. Cane sugar is very readily hydrolysed in comparison with maltose, cellobiose and the like, and so too are polysaccharides like the inulin of Compositae and the levans of grasses, which are polymers of fructofuranose.

In case any apology is needed for this chemical diversion from our main topic, it can truthfully be said that, since in biochemical circles we always talk and think about sugars in these ring forms in preference to the older and spatially less real forms, such a diversion is really necessary in case any of those who read this book should be tempted to dig a little deeper into the biochemical literature.

A last word about two pentoses of great biochemical importance, namely, *ribose* and *deoxyribose*, the formulae of which are the following:

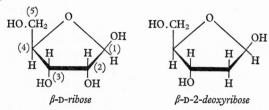

β-D-*ribose* β-D-2-*deoxyribose*

The difference between them lies in the absence of an O atom at carbon 2 in deoxyribose, which can more descriptively be called D-2-deoxyribose. Both sugars are important as constituents of nucleic acids, to which a separate chapter is devoted later on, and are mentioned here for that reason. In biological combination, like biologically combined fructose, they occur in the five-membered forms pictured above.

64

8

PHOTOSYNTHESIS OF CARBOHYDRATES

1. INTRODUCTION

Even a cursory study of any natural habitat reveals the fact that animals are intimately dependent upon green plants. Herbivorous animals use plant materials at first hand, carnivores at second or third hand, and there are many reasons for this. Given light, water and atmospheric carbon dioxide the green plants can produce carbohydrates and fats in abundance. Given also an appropriate source of nitrogen they can synthesize amino acids and proteins. Quite apart from all these, however, plants produce an immense variety of other substances, ranging from such exotic compounds as morphine, strychnine, the digitalis alkaloids and many more, to a variety of simpler substances which play parts of vital importance in plant metabolism. Most of this latter group are as important in animal metabolism as they are in that of plants but many of them, known as vitamins, animals cannot synthesize for themselves. Once again, therefore, animals must go back to the green plants. In the end, then, animals rely upon plants to provide them with fuel, essential amino acids and vitamins at the very least.

The driving force behind all this is, of course, solar radiation. With its aid the green plants synthesize everything they need to maintain and reproduce themselves, with enough left over to support all the animal life on this planet. Obviously then we must consider the phenomenon of photosynthesis.

2. THE ACTIVATION OF CHLOROPHYLL

In any undisturbed population of molecules the majority will be in what is called the *ground state*. The electrons circulating in their orbital paths will be as close to the nucleus as atomic structure permits.

Now there are many molecules—and the only ones we need consider here are those of chlorophyll—which on exposure to light can absorb photons and pass into an activated or excited state. The energy of one photon is handed over to or incorporated into that of one electron, which flies off into a new orbit more remote from the nucleus. Later on this electron falls back into its normal orbit, shedding its extra energy as it does so. Often this energy appears in the form of fluorescent light; in fact a solution of chlorophyll, bright green in colour, displays a strong red fluorescence on exposure to light. Evidently, the primary function of light in photosynthesis consists in the activation or excitation of chlorophyll molecules.

3. ACTIVATION IN CHLOROPLASTS

Several characteristically biochemical changes accompany the photo-chemical activation of chlorophyll and its subsequent return to the ground state, but these are only observable in *intact chloroplasts*. These are the microscopic, intracellular organelles in which the chlorophyll is contained.

In a series of very elegant experiments by R. Hill it was shown that a suspension of isolated, illuminated chloroplasts will reduce ferric oxalate to the ferrous state, evolving oxygen at the same time, although no fixation of carbon dioxide takes place. Out of these seemingly simple observations a number of most important discoveries have emerged. First of all, any of a wide range of biological substances, including a group of most important coenzymes, can replace ferric oxalate and undergo reduction in its place.

In other experiments the water used was enriched with heavy oxygen (^{18}O) and the isotope turned up in the oxygen which was evolved. Thus, for the first time, it was established that this oxygen comes from water, not from CO_2 as was formerly supposed. Clearly water can undergo *photolysis* giving molecular oxygen and (presumably nascent) hydrogen. The latter could well be responsible for the 'reducing atmosphere' whereby many biological compounds undergo reduction instead of the oxidation they normally suffer in the 'oxidizing atmosphere' of animal tissues. There is something of a parallel here in the electrolysis of water which takes place in a car battery on charge.

Nor is this all. Whereas activated chlorophyll in solution sheds much of its surplus energy in the form of fluorescence on returning to the ground state, in intact chloroplasts a large part of the surplus energy is returned to the environment in the form of *adenosine triphosphate* (ATP), formed from adenosine diphosphate (ADP). Later on we shall have much more to say about these two compounds; for the moment suffice it to say that the conversion of ADP into ATP constitutes a mechanism for storing energy, while the conversion of ATP to ADP can provide energy for a great variety of biological processes, from muscular contraction to chemical synthesis. What is interesting in our present context is that ATP is required at several stages in the reactions that lead on to the synthesis of starch.

In isolated chloroplasts, then, three things happen: (i) Creation of a 'reducing atmosphere'; (ii) Photolysis of water and evolution of oxygen; (iii) Provision of ATP for later stages in the formation of starch.

4. FIXATION OF CARBON DIOXIDE

We have glanced briefly at a few observations on chlorophyll in simple solution and on intact chloroplasts but nothing has yet been said about CO_2-fixation or carbohydrate synthesis. Neither of these phenomena can be observed in suspensions of chloroplasts, still less in solutions of chlorophyll. To study these processes *intact cells* are necessary and most of the work in this field has been done with suspensions of microscopic unicellular algae such as *Chlorella*.

In barest outline the procedure is this. Thick suspensions of *Chlorella* are put up in a suitable medium containing radioactive bicarbonate (^{14}C) as a source of CO_2. The suspensions are briefly illuminated, killed as rapidly as possible, extracted and analysed by paper chromatography. Radio-autographs are printed from the chromatograms and show into which spot or spots radioactive carbon has been incorporated.

After exposure to light for a few seconds radioactivity is already detectable in many sugars and sugar derivatives and even in a few amino acids. By reducing the period of illumination however a point is eventually reached at which only one spot shows any radioactivity. This spot corresponds to *glyceric acid phosphate*

((P) corresponds to a phosphate residue)

$$
\begin{array}{l}
CH_2O(P) \\
| \\
CHOH \\
| \\
COOH
\end{array}
$$

This, then, is the first detectable product of CO_2 fixation.

Now *two* molecules of glyceric acid phosphate are formed for each molecule of CO_2 fixed. This can readily be explained if we assume that there is a 5-carbon acceptor molecule of some kind which adds on CO_2 to form an unstable 6-carbon intermediate, which then splits into 2 molecules of the 3-carbon acid phosphate. Such a 5-carbon acceptor substance does indeed exist in, and has been isolated from, green plants in the shape of *ribulose*-1 : 5-*diphosphate* so that, thus far, the photosynthetic pathway is as follows:

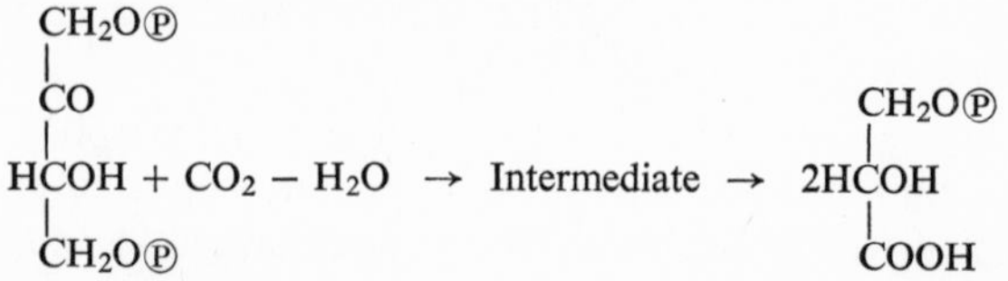

$$
\begin{array}{c}
\text{CH}_2\text{O}\textcircled{P} \\
|\\
\text{CO} \\
|\\
\text{HCOH} \\
|\\
\text{CH}_2\text{O}\textcircled{P}
\end{array}
+ \text{CO}_2 - \text{H}_2\text{O} \;\rightarrow\; \text{Intermediate} \;\rightarrow\; 2
\begin{array}{c}
\text{CH}_2\text{O}\textcircled{P} \\
|\\
\text{HCOH} \\
|\\
\text{COOH}
\end{array}
$$

   ribulose-1:5-*diphosphate*   *glyceric acid phosphate.*

Now, there exists in the *green parts of plants*, but not elsewhere, a triose phosphate dehydrogenase which, like most dehydrogenases, can act reversibly, but unlike all other known triosephosphate dehydrogenases, which are numerous and very widely distributed and require a coenzyme commonly known as NAD*, this plant enzyme has a special and specific requirement for a different co-factor, NADP†.

Given glyceric acid phosphate and the reduced form of NADP, i.e. NADP.H_2, the following reaction takes place and requires ATP to drive it towards completion:

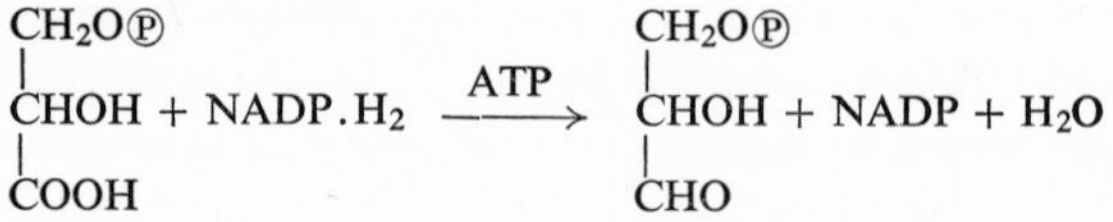

$$
\begin{array}{c}
\text{CH}_2\text{O}\textcircled{P} \\
|\\
\text{CHOH} \\
|\\
\text{COOH}
\end{array}
+ \text{NADP.H}_2 \xrightarrow{\;\text{ATP}\;}
\begin{array}{c}
\text{CH}_2\text{O}\textcircled{P} \\
|\\
\text{CHOH} \\
|\\
\text{CHO}
\end{array}
+ \text{NADP} + \text{H}_2\text{O}
$$

   glyceric acid phosphate   *glyceraldehyde phosphate*

Since light is required to provide the ATP and the reduced coenzyme the reactions up to this point are light dependent, but all the subsequent operations can take place in the dark.

The next reaction, isomerisation of glyceraldehyde phosphate, is catalysed by an *isomerase* already mentioned in these pages (p. 44):

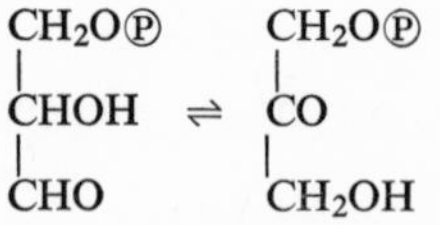

$$
\begin{array}{c}
\text{CH}_2\text{O}\textcircled{P} \\
|\\
\text{CHOH} \\
|\\
\text{CHO}
\end{array}
\;\rightleftharpoons\;
\begin{array}{c}
\text{CH}_2\text{O}\textcircled{P} \\
|\\
\text{CO} \\
|\\
\text{CH}_2\text{OH}
\end{array}
$$

        dihydroxyacetone phosphate

One further reaction leads to fructose-1 : 6-diphosphate, a substance that lies directly on the pathway leading to starch formation and we shall take up this part of the story in the next chapter. The reaction

  * NAD = nicotinamide adenine dinucleotide.
  † NADP is a phosphorylated derivative of NAD.

which produces this progenitor of starch is catalysed by a lyase, *aldolase,* and proceeds as follows:

$$
\begin{array}{lcl}
CH_2O\textcircled{P} & & CH_2O\textcircled{P} \\
| & & | \\
CO & & CO \\
| & & | \\
CH_2OH & & CHOH \\
+ & & | \\
CHO & \longrightarrow & CHOH \;\;-------\rightarrow \text{ starch} \\
| & & | \\
CHOH & & CHOH \\
| & & | \\
CH_2O\textcircled{P} & & CH_2O\textcircled{P} \\
& & \textit{fructose-}1\!:\!6\textit{-diphosphate}
\end{array}
$$

Thus, according to the scheme as we have delineated it here, *one molecule of ribulose diphosphate is consumed for each* CO_2 molecule fixed.

It might be anticipated that polysaccharide synthesis would cease as soon as the plant's stock of ribulose diphosphate is exhausted but, in point of fact, ribulose diphosphate is continually being regenerated by a complex network of subsidiary reactions, starting with triose phosphate. What is more, these accessory reactions are coupled to the processes we have already considered and coupled, moreover, in such a manner that for every molecule of fructose diphosphate entering into starch formation, *a new molecule of ribulose diphosphate becomes available.* Here again, as is so often the case in biochemical operations, we have a system which, considered in its entirety, is essentially cyclical in nature.

9

FUEL FOR THE MACHINE: CARBOHYDRATE

DIGESTION OF CARBOHYDRATES

Of the three major foodstuffs which together make up the bulk of an average diet, proteins, as we have seen, are specially concerned with the formation and maintenance of the structural and catalytic machinery of the living cell. If more protein is consumed than is required to fulfil these functions the surplus amino acids are de-aminated and the nitrogen-free residues are used to supplement the stocks of fat and carbohydrate. These are the substances which, above all, serve as fuel for the biochemical machine.

The chief *monosaccharides* found in common foodstuffs are glucose and fructose. These require no digestion and are absorbed as such. There are, however, three common *disaccharides* which do require digestion, namely maltose (malt sugar), lactose (milk sugar) and sucrose (cane or beet sugar). Digestion of these takes place in the small intestine, the intestinal juice containing three hydrolytic disaccharases, one specific for maltose (maltase), another for lactose (lactase) and a third for sucrose (saccharase, sucrase or invertase). These catalyse the hydrolysis of their respective substrates to give the component monosaccharides:

maltose → glucose + glucose,
lactose → glucose + galactose,
sucrose → glucose + fructose.

The chief *polysaccharide* of interest is starch. In the uncooked form starch is relatively indigestible, but cooking breaks up the natural grains in which starch is usually found and renders the starch itself accessible to the hydrolases which digest it. Two such enzymes are concerned, both called *amylases*. The first of these occurs in saliva (salivary amylase) and the second in pancreatic juice (pancreatic amylase). These catalyse the hydrolysis of starch but not, as might have been anticipated, with the liberation of single glucose molecules but with that of pairs of glucose molecules in the form of *maltose*. This maltose is later hydrolysed by the intestinal maltase, yielding *glucose* which, along with other monosaccharides, is absorbed from the small intestine.

The first port of call of blood leaving the small intestine is the liver, and this organ contains *isomerizing enzymes* whereby fructose and the less common monosaccharides, galactose and mannose, can all be converted into glucose. This the liver takes up and polymerizes to form the characteristic animal polysaccharide, *glycogen*, which is held in stock for distribution to other parts of the body as and when they need it.

Needless to say, food, especially plant food, contains a good deal of polysaccharide materials of various kinds, quite apart from starch. Of these *cellulose* is far and away the most abundant; in fact it is probably the commonest polysaccharide on the face of the earth. Although, like starch, it is composed entirely of glucose units, the units are linked in the β-fashion instead of the α-fashion found in starch and so it is not attacked by the amylases. It is very insoluble indeed and very few animals seem to have evolved enzymes capable of splitting it into its component parts. Even herbivorous animals like the cow and the sheep, though they get a great deal of nourishment out of cellulose, cannot digest it for themselves but delegate the job to hordes of micro-organisms which live in their stomachs. Chewing the cud involves a thorough mixing of the food with these micro-organisms and the mixture is subsequently swallowed and incubated in specialized digestive organs under conditions ideal for bacterial activity. The micro-organisms break up cellulose to produce, not free glucose or any other sugar, but large volumes of methane and hydrogen, a somewhat explosive mixture which is useless as far as the cow is concerned, together with high yields of *acetic, propionic and butyric acids*. These the cow absorbs and stores away in the form of carbohydrate and fat. Much the same is true of most herbivorous animals from cows to cockroaches; always there is, somewhere in the intestinal canal, a large microbial population and a capacious dilatation in which the food and the micro-organisms can be incubated together. Animal cellulases are rare indeed, but there are a few authentic cases, and among these the best known is the cellulase of the edible snail. A recent addition to the list is that of the silverfish.

STORAGE OF CARBOHYDRATES

Carbohydrate is stored in plants mainly as starch and a variety of fructose polymers, and in animals as glycogen. The mechanisms

Fuel for the Machine: Carbohydrate

underlying the synthesis of glycogen and starch are well understood and are essentially similar. Although both are formed from glucose units, neither of these polysaccharides can be directly synthesized, either in the laboratory or in the living cell, from free, unmodified glucose. In fact it seems that *glucose itself can neither be stored nor metabolized unless it is first converted into a phosphate ester.*

Two glucose phosphates are important in this connexion, the first, glucose-6-phosphate, being produced by an irreversible reaction between glucose and a phosphate donor known as *adenosine triphosphate* (ATP):

$$\text{glucose} + \text{ATP} \rightarrow \text{glucose-6-phosphate} + \text{ADP}.$$

Here the enzyme, a member of the group of transferases, is called *hexokinase*; ADP is adenosine *di*phosphate. An isomerizing enzyme, *phosphoglucomutase*, now catalyses the reversible conversion of glucose-6-phosphate into its isomer α-glucose-1-phosphate:

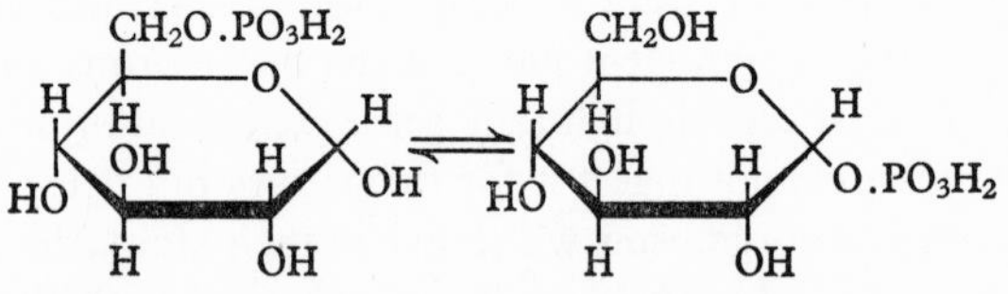

glucose-6-phosphate   *α-glucose-1-phosphate*

Starch and glycogen can be then formed by a series of reactions in each of which a fresh glucose unit is transferred from phosphate to an already partly formed 'starter' by elimination of phosphoric acid. The enzyme concerned is a *transglucosylase*:

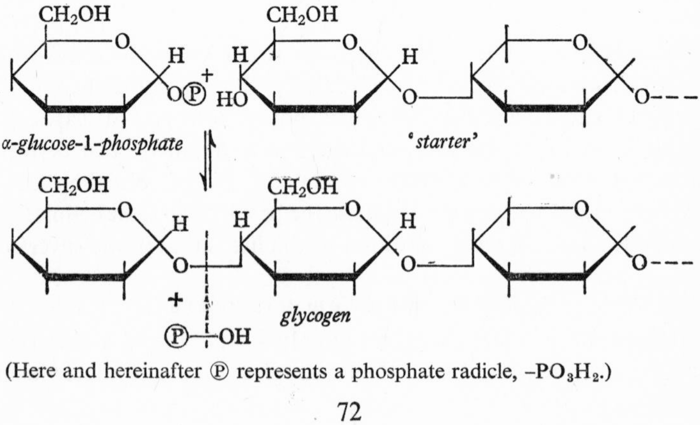

α-glucose-1-phosphate   'starter'

glycogen

(Here and hereinafter ℗ represents a phosphate radicle, $-PO_3H_2$.)

72

Storage of Carbohydrate

In this way a long chain of glucose units can be built up so that the product formed resembles amylose (p. 57). The branching arrangement of chains found in amylopectin and glycogen (p. 58) is produced later on by the action of so-called 'branching enzymes' which remove pieces of the amylose chain and transplant them to other points in the main chain itself.

It will be observed that in the foregoing equations, once glucose-6-phosphate has been formed, the rest of the reactions leading to polymerization are reversible, and it is in fact possible to produce amylose-like products experimentally if we start with α-glucose-1-phosphate together with the transglucosylase and a trace of pre-formed glycogen or amylose to act as the 'starter'. However, the overall equilibrium conditions are much in favour of the glucose phosphates, so that it is necessary for the experimental demonstration of this polimerization to start with very high concentrations of α-glucose-1-phosphate, concentrations far in excess of those likely to be encountered in biological systems. Indeed the reactions described so far are used biologically for the breakdown rather than for the synthesis of these important polysaccharides: their synthesis calls for different reactions, the result of which is to shift the equilibrium in the direction of polysaccharide rather than sugar phosphate.

This additional reaction involves a compound called *uridine triphosphate* (UTP), which is rather closely related to adenosine triphosphate (ATP). Instead of handing over a glucose unit directly to the 'starter', the α-glucose-1-phosphate reacts first with UTP to form *uridinediphosphateglucose* (UDPG) and it is from this that the glucose unit is transferred to the 'starter'. Uridine diphosphate is set free, to be reconverted into UTP by reacting with ATP and used over again.

It may be asked why, when the reaction proceeds in this rather more complex fashion, the equilibrium between the sugar phosphates and the polymer should be changed. The reason has to do with the energy content of the substances concerned. It will be remembered that free glucose cannot be converted into the polymer until after it has been transformed into glucose-6-phosphate by the hexokinase reaction with ATP. For reasons that will be considered in a later chapter, the transfer of the phosphate group from ATP to glucose is accompanied by the transfer of a considerable amount of energy so that, in effect, glucose-6-phosphate is a more reactive substance than the original free glucose. The later reaction of α-glucose-1-

73

phosphate with UTP provides an additional 'push' in the direction of polysaccharide formation. It seems almost certain that many polysaccharides other than glycogen and starch also involve reactions in which UTP plays a similar part, and UDP-sugar complexes are known too to be involved in the synthesis of many disaccharides such as sucrose and lactose.

These are only some of the many synthetic processes going on in the body in which energy drawn from ATP is used to push reactions in the forward, i.e. in the synthetic, direction. Sometimes ATP is used directly, as in the hexokinase reaction: less commonly it is used indirectly, for example in the rephosphorylation of UDP after UDPG has transferred its glucose residue in the synthesis of polyglucoses.

Although this latter is a rather intricate performance it is described here in some detail because it is fairly typical of the biological mode of synthesis of large molecules and also because it can help us to understand how glycogen is formed from dietary monosaccharides, and how supplies of glucose are subsequently transmitted, through the blood, to the other tissues of the animal body.

This example is important too because it illustrates a very general biochemical phenomenon for it is very usually found that the breakdown and synthesis of any given substance usually proceed along different pathways, catalysed by different enzymes, rather than by simple reversal of a single enzyme-catalysed reaction. Another simple example consists in the interconversion of glucose and glucose-6-phosphate which can be represented thus:

$$\begin{matrix} H_3PO_4 \\ H_2O \end{matrix} \Big\rangle\!\!\Big\langle \begin{matrix} \text{glucose} \\ \text{glucose} \quad 6\textcircled{P} \end{matrix} \Big\rangle\!\!\Big\langle \begin{matrix} \text{ATP} \\ \text{ADP} \end{matrix}$$

Both steps are irreversible for all practical purposes.

DISTRIBUTION OF CARBOHYDRATE

First, liver glycogen undergoes a process often known as *phosphorolysis*, since HO℗ is involved, i.e. the removal, one at a time, of glucose units. As we have seen, these come off in the form of α-glucose-1-phosphate and the reaction is catalysed by the liver transglucosylase, α-glucan phosphorylase. Phosphoglucomutase, an isomerase, con-

74

verts the product into glucose-6-phosphate, and the latter is then hydrolytically split by glucose-6-phosphatase to give free glucose:

glycogen → α-glucose-1-phosphate → glucose-6-phosphate → glucose + H_3PO_4

The α-glucose-1-phosphate is not attacked by this enzyme, which is specific for the 6-ester.

The free glucose enters the blood stream and is carried away to other tissues, all of which, apart from the brain, contain hexokinases and ATP and the other enzymes necessary to lay down local stocks of glycogen through essentially the same processes as the synthetic reactions that go on in the liver.

In the ordinary way the rate at which glycogen is broken down in the liver is nicely balanced against the rate at which it is taken up by the other tissues, so that the concentration of glucose in the blood remains very constant at approximately 100 mg/10 ml—another example of internal constancy. In this case, however, no great constancy is found except in warm-blooded animals, and especially in the mammals, which are characterized by the extremely high development of their brain and nervous system generally. Unlike the other tissues, the brain has no local store of glycogen. It metabolizes very little apart from glucose and is therefore intimately dependent upon glucose brought to it by the blood. The importance of a constant level of blood sugar is therefore particularly great in these higher animals. A fall in blood sugar means that the brain is starved of this metabolite and reacts by firing off uncontrolled and unco-ordinated impulses to the body generally, producing convulsions. An elevation of the blood sugar in itself does not have any such adverse effects, but elevation does lead to a loss of glucose by way of the kidneys and is therefore very wasteful of this important energy-yielding substance.

The normal level of blood sugar is maintained by a balance between a number of *hormones*, i.e. internal secretions produced by specialized glandular tissues and liberated directly into the blood stream. Among these hormones two are of particular importance in the control of carbohydrate metabolism. Insulin, which is formed by special so-called 'islet cells' in the pancreas, tends to lower the blood sugar level and sets a higher limit. The second, which may be called an 'anti-insulin hormone', is formed by the anterior lobe of the pituitary gland, and sets a lower limit. If either of these two limits is passed the level is corrected, in a normal animal, by the production of more of the hormone setting that limit, so that the glucose level is forced back to within its normal physiological range. An example of this can be seen if an animal is given a large dose of glucose by mouth. The blood sugar level begins to rise as the glucose is absorbed, but if the rise is too large or too fast the level begins to fall again, even though glucose is still being absorbed from the intestine. This is due to a sudden burst of insulin formation on the part of the islet cells of the pancreas, the insulin forcing the blood glucose level downwards towards its normal value. Sometimes the level may be temporarily pushed down even below its usual lower limit.

It is important to realize that the level of the blood sugar is regulated by the *balance* between these internal secretions rather than by the absolute amount of either. If this balance is upset through either under- or overproduction of one or other of the two regulating factors, disease results. The condition known as diabetes (diabetes mellitus) is produced in this way. Here disease or disorder of the pancreatic islet tissue leads to underproduction of insulin, so that the influence of the anti-insulin hormone gets the upper hand. The blood-sugar level rises, partly because of removal of the influence of insulin and partly because, at the same time, less glucose than usual is metabolized, so that one of the signs of diabetes is an abnormally high level of glucose in the blood. Because of the high blood level, glucose leaks away through the kidneys and into the urine, and a second sign is the presence of free glucose in the urine.

This particular aspect of diabetes is not in itself particularly serious, apart from its wastefulness, and the often fatal outcome of untreated cases of diabetes is due to something else. We have noticed that *carbohydrate metabolism is subnormal in the diabetic*. But the organism must still get its energy supplies from somewhere, and does so by metabolizing abnormally large amounts of fat. In diabetes even the metabolism of this fat is incomplete. *Intermediary ketonic products arising from fat* begin to accumulate in the blood, and make their appearance in the urine. Being acids, they tend to force the pH of the blood towards the acid side and, being ketones at the same time poison the patient, who sinks into a diabetic coma. All these can be corrected by the administration of insulin in proper quantities but, of course, one or two doses do not suffice to cure the disease; regular and appropriate injections of insulin are required to control it.

Distribution of Carbohydrate

METABOLISM OF GLUCOSE AND GLYCOGEN

The breakdown of glucose and glycogen involves a long procession of ordered and orderly chemical events. We shall not attempt to deal with them here except in the barest outline. Apart from glucose and glycogen themselves, all the intermediates are phosphorylated compounds, and the first few stages can be represented thus:

$$\text{starch or glycogen} \rightleftharpoons \alpha\text{-glucose-1-phosphate}$$
$$\text{glucose} \xrightarrow{\text{ATP}} \text{glucose-6-phosphate}$$
$$\updownarrow$$
$$\text{fructose-6-phosphate}$$
$$\updownarrow \text{ ATP}$$
$$\text{fructose-1-6-diphosphate}$$

It must be emphasized at this point that all the reactions leading from fructose diphosphate back to glycogen or starch as the case may be can be operated in reverse, and it is at the fructose diphosphate stage (p. 68) that the *photosynthetic reaction sequence* joins the main line of breakdown and synthesis of polysaccharide material.

On the breakdown pathway the fructose diphosphate next undergoes splitting by a lyase, *aldolase*, to give phosphate esters of the two triose phosphates, i.e. the phosphates of glyceraldehyde and dihydroxyacetone (see p. 44). These are further manipulated through a long series of reactions to give, eventually *pyruvic acid*. Even this is by no means the end of the road, which leads in the end to CO_2 and water, but is a convenient point at which to pause for a while and look further before we leap on to one of the most exciting and remarkable discoveries of biochemistry of all time, the *citric acid cycle*.

Pyruvic acid is an α-keto acid and it can, in common with other α-keto acids, undergo a reaction known as *oxidative decarboxylation*, which can be represented in oversimplified form thus:

$$CH_3COCOOH + H_2O - 2H \rightarrow CH_3COOH + CO_2.$$

The importance of this reaction is very great indeed, for not only does carbohydrate pass through the stage of pyruvate but, as will be remembered, α-keto acids are formed by the deamination of the amino acids (p. 48) and these, like pyruvate, also undergo oxidative decarboxylation. This process is, in reality, a very complex one, involving six or more enzymes and at least five co-factors, one of which we cannot avoid mentioning, namely *coenzyme* A. This is a large and very complicated molecule and the reader can be spared a great deal of pain if we refer to it simply as CoA.SH. CoA.SH plays a part in the detachment of CO_2 from the α-keto acid and appears in an intermediate product of that reaction, viz. *acetyl-S-CoA*. We shall return to discuss its fate later on.

A word of explanation is needed to show why coenzyme A is usually abbreviated as CoA.SH. The coenzyme possesses a thiol (−SH) group and when it enters into biological combination—in the formation of acetyl-coenzyme A for example—it does so through this −SH radical. This can be shown by allowing the coenzyme to react with iodoacetate;

$$-SH + I.CH_2COOH \rightarrow -S.CH_2COOH + HI.$$

This blocks the functional −SH group and the coenzyme becomes totally inert.

10

FUEL FOR THE MACHINE: FAT

CHEMICAL NATURE OF FATS

Although many of the fatty substances occurring in nature are very complex, the most abundant and those with which we shall deal here are among the most simple and are, indeed, often referred to as *simple fats*. These substances are tri-esters of the trihydric alcohol, *glycerol* (glycerine) and are important fuels by the oxidation of which living cells obtain a large proportion of the energy they need. Usually the cell oxidizes a nicely balanced mixture of carbohydrate and triglyceride material for energy production. A typical simple fat can be formulated thus:

$$CH_2O.OC.R_1$$
$$CHO.OC.R_2$$
$$CH_2O.OC.R_3$$

Here R_1, R_2 and R_3 correspond to *fatty acid chains*, among the most common of which are the saturated long-chain acids stearic (18 C), palmitic (16 C) and the unsaturated oleic acid (18 C) with one double bond in the middle of the chain. The naturally occurring fatty acids always contain an *even number of carbon atoms*. Generally speaking, the larger the proportion of unsaturated fatty acids present, the lower is the melting or softening point of the fat. Thus, while beef and mutton fat are mainly made up of saturated acids and are solid at ordinary temperatures, olive oil, which contains little but oleic acid, is liquid, though not so liquid that it will not freeze if kept in a refrigerator. Butter fat contains a fair proportion of the 4-carbon butyric acid and has a soft consistency.

In some cases natural fats, known as *waxes* because of their consistency, are esters of unusually long-chain alcohols containing 22, 24 or even more carbon atoms, and most animal waxes such as beeswax and spermaceti are built up in this way.

Although the three fatty acids present in a typical fat are sometimes all of the same kind—olive oil, for example, has three molecules of oleic acid for each molecule of glycerol—this is the exception rather than the rule. In a given sample of animal fat any individual molecule may contain one, two or three different fatty acid residues and one

Fuel for the Machine: Fat

molecule may differ from another in its fatty acids, so that animal fats generally are somewhat complicated mixtures of molecular species.

DIGESTION OF FATS

The first stage in the digestion of fats is catalysed by hydrolytic enzymes known as *lipases*. One interesting characteristic of these enzymes is that their action is freely reversible and the behaviour of the castor oil bean (*Ricinus*) lipase is illustrated in Fig. 10 (p. 43). The figure shows how, whether the enzyme acts upon triolein or upon its constituents, the reaction ends up with an equilibrium mixture of the ester, the alcohol and the acid. However this is not to say that natural fats are synthesized in this way; like the polysaccharides, fats are broken down and synthesized along different pathways, catalysed by different enzymes.

The main supply of digestive lipases in the animal body is secreted by the pancreas, and digestion and absorption of the products are facilitated by bile, secreted by the liver. Pancreatic lipase does not necessarily remove all three of the R groups simultaneously but one after another, so that after a short period of digestion the mixture contains mono-, di- and undigested triglycerides, together with some free fatty acids and glycerol. Now the bile contains substances known as *bile salts*, the chief characteristic of which is their power to lower the surface tension at fat/water interfaces to a very low level. In the presence of bile salts and the products of partial digestion, the undigested fatty material becomes so finely emulsified that the resulting particles can pass unchanged through micropores in the wall of the intestine. The absorbed particles are carried away in the form of microscopic droplets to be deposited in the fat depots of the body; for example, in the subcutaneous tissues and the mesenteries. It is estimated that two-thirds to three-quarters of the total dietary fat is absorbed in the emulsified form. The absorption of those parts of the total fat that have been digestively dismantled raises other problems which will not be considered here.

METABOLISM OF FATS

The first stage in the metabolism of fats takes place in the liver and consists in the splitting of triglycerides into glycerol and the constituent fatty acids by means of lipases. In all probability, glycerol is transformed into carbohydrate, while fatty acids undergo oxidation, beginning at the carboxylic group and leading to CO_2 and water.

Metabolism of Fats

Elucidation of the mechanisms involved in the oxidation of the fatty acids began many years ago with the work of Knoop, who 'labelled' short fatty acid chains in the ω-position, i.e. at the carbon atom most remote from the acidic group, by the introduction of a phenyl group ($-C_6H_5$). When these substances were administered to dogs, correspondingly labelled products were found in the urine. This was one of the earliest investigations in which use was made of labelling technique, a method which has been enormously extended since heavy and radioactive isotopes have become easily available, if still very expensive. Moreover, the use of isotopes as labels has one great advantage over purely chemical labelling. Whereas it is always possible that the introduction of an abnormal chemical label, as in Knoop's experiments, might provoke new and abnormal metabolic reactions, nearly all isotopes, by contrast, are chemically indistinguishable. Equally they are metabolically indistinguishable and can therefore be used with reasonable certainty that metabolic processes discovered or elucidated with their aid are identical with those normally taking place.

It might have been expected that fatty acid chains would lose one carbon atom after another until the whole chain had been eaten away. As often happens, however, expectation proves at variance with facts. It turned out, in fact, that when the parent ω-phenylated fatty acid was one with an even number of carbon atoms, the end-product was *phenylacetic acid*, e.g.

$$C_6H_5.CH_2CH_2 \vdots CH_2COOH \rightarrow C_6H_5.CH_2COOH.$$

If, on the other hand, the parent acid was one with an odd number of carbon atoms the end-product was *benzoic acid*, e.g.

$$C_6H_5.CH_2 \vdots CH_2CH_2 \vdots CH_2COOH \rightarrow C_6H_5.COOH.$$

If carbon atoms could be removed one at a time it would be expected that all ω-phenylated fatty acids would give rise to benzoic acid: even phenylacetic acid would behave in this way. It follows beyond reasonable doubt that the carbon atoms cannot be detached singly but *must be removed in pairs*. This might be accomplished by the following series of reactions;

$$\overset{\beta}{}\quad\overset{\alpha}{}$$
$$......CH_2CH_2CH_2COOH$$
$$\Big\downarrow \quad oxidation\ at\ \beta\text{-}carbon$$
$$......CH_2CO.CH_2COOH$$
$$+H_2O \Big| \quad hydrolytic\ split$$
$$......CH_2COOH + CH_3COOH$$

81

Fuel for the Machine: Fat

Since the naturally occurring fatty acids contain always an even number of carbon atoms, the long chain acids could be wholly degraded into 2-carbon units by successive removal of pairs of C atoms. Much new evidence has now been gathered in support of this hypothesis of *β-oxidation* and today there can be no doubt that carbon atoms are removed from fatty acid chains in pairs, though the reaction sequence originally postulated is now known to be only a part of the whole truth.

In the last paragraph one oversimplification has been used and attention may be drawn to it before we go further. Benzoic and phenylacetic acids do not appear in the urine as such but in the form of compounds of these substances with the simple amino acid *glycine*;

$$C_6H_5COOH + H_2N.CH_2COOH \longrightarrow C_6H_5CO.HN.CH_2COOH$$
$$\text{\textit{hippuric acid}}$$

$$C_6H_5CH_2COOH + H_2N.CH_2COOH \longrightarrow C_6H_5CH_2CO.HN.CH_2COOH$$
$$\text{\textit{phenaceturic acid}}$$

The interesting point here is that, unlike the parent acids, which are rather toxic substances, *the glycine derivatives are relatively innocuous*. Many comparable cases are now known; cases in which the administration of substances toxic to the organism is followed by chemical manipulations which render them innocuous or relatively so. To take another example, aniline, a highly poisonous compound, is eliminated in the form of its acetyl derivative, acetanilide, which is itself sufficiently harmless to be used extensively as an antipyretic (fever-reducing) drug. Indeed, it was so used for many years prior to the discovery of aspirin, an even less toxic antipyretic.

Discoveries of this kind led to the introduction of the term '*protective synthesis*', but this is something of a misnomer because cases are known in which substances harmless in themselves undergo metabolic conversion into highly poisonous compounds when administered to animals. The truth, evidently, is this; that quite apart from the enormous number and variety of metabolic reactions that can be accomplished by cells acting upon normal metabolites, enzymes can sometimes act also upon certain foreign substances and produce abnormal materials from them. Whether the products are more or less toxic than the parent compounds seems to be largely fortuitous.

Going back now to the original theory of *β*-oxidation it will be seen that each pair of carbon atoms is visualized as being set free in

82

the form of acetic acid. One of the major obstacles to the theory was the fact that no acetic acid could be detected, even when intense fatty acid breakdown was taking place. In many cases, moreover, aceto-acetic acid (CH_3COCH_2COOH) was found to be formed, especially in experiments in which little or no simultaneous carbohydrate breakdown was taking place. In liver perfusion experiments, for example, fatty acid breakdown gives rise to a large-scale production of acetoacetate, especially if the liver has been taken from an animal previously starved so that its reserves of glycogen have been depleted. Evidently, therefore, *fat metabolism is linked in some way to that of carbohydrate.*

In the ordinary way, in an intact, normal animal for instance, where carbohydrate and fat metabolism go hand in hand, only the merest traces of acetoacetate formation can be detected. Such aceto-acetate as is formed in the liver is carried away to the peripheral tissues and there oxidized. If, however, an animal is depleted of carbohydrate, whether because it is diabetic or has been starved of carbohydrate, in all these cases and in many others in which carbo-hydrate metabolism is subnormal, *acetoacetate* begins to be formed on a large scale. From it two other substances arise, viz. *β-hydroxy-butyric acid*, formed by an oxidoreductase, and *acetone*, formed by an apparently spontaneous decarboxylation of acetoacetate;

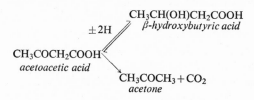

$$CH_3CH(OH)CH_2COOH$$
β-hydroxybutyric acid

$\pm 2H$

$$CH_3COCH_2COOH$$
acetoacetic acid

$$CH_3COCH_3 + CO_2$$
acetone

Collectively known as the ketone or acetone bodies, these are the really toxic agents in untreated cases of diabetes (see p. 76).

Let us now take into account the results of some more recent work on fatty acid oxidation. The reader may recall the important part played by *coenzyme* A in the oxidation of carbohydrate by way of pyruvate (p. 77). It has now been found that $CoA.SH$ is important in fatty acid oxidation also, and at every stage of the $β$-oxidative process. The reactions are summarized below. First of all, fatty acids undergo no oxidative change unless they are first converted into their acyl-S.CoA derivatives. Then, and only then, the $β$-oxidative process begins, leading by way of several intermediate reactions to the formation of the β-keto-acyl fatty acyl-S.CoA derivatives. The splitting of these compounds, attended by the removal of two carbon fragments, is not hydrolytic as was formerly believed, but thiolytic. The splitting agent is not water but CoA.SH, the term thiolytic being used with reference to the fact that CoA.SH is a sulphur (thiol) compound.

Compare the reactions as we now know them to take place with those originally postulated (p. 81):

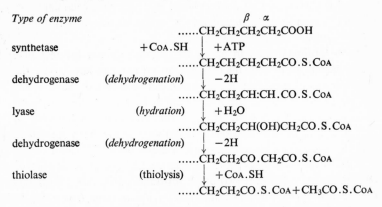

Type of enzyme		β $\quad$ α
		......$CH_2CH_2CH_2CH_2COOH$
synthetase	$+CoA.SH$	$+ATP$
		......$CH_2CH_2CH_2CH_2CO.S.CoA$
dehydrogenase	(*dehydrogenation*)	$-2H$
		......$CH_2CH_2CH:CH.CO.S.CoA$
lyase	(*hydration*)	$+H_2O$
		......$CH_2CH_2CH(OH)CH_2CO.S.CoA$
dehydrogenase	(*dehydrogenation*)	$-2H$
		......$CH_2CH_2CO.CH_2CO.S.CoA$
thiolase	(thiolysis)	$+CoA.SH$
		......$CH_2CH_2CO.S.CoA+CH_3CO.S.CoA$

This is a most elegant series of reactions. Once the fatty acid has been prepared for degradation by its union with CoA.SH, oxidative degradation can begin under the influence of the appropriate enzymes, most of which have now been highly purified. Most interesting of all is the thiolytic reaction, for the residual fatty acid emerges already united with CoA.SH and ready for the next oxidative step, and the two-carbon product is acetyl–S CoA.

Acetyl–S.CoA is known to be a highly reactive substance in cells and tissues, and this explains why, in earlier times, it was not possible to detect the free acetic acid required by the older theory. Knoop himself had some inkling of this, for he attributed the difficulty of demonstrating the formation of acetate to the possibility that it might be split off in the form of some highly reactive derivative ('active acetate') which was carried off by other reactions as fast as it was produced.

If the reader will look again at the last part of chapter 9, he will be reminded that carbohydrate too undergoes breakdown to acetyl–S.CoA, so that these two main and independent lines of breakdown converge towards a common product. Further, of the amino

Metabolism of Fats

acids there are some that give rise to pyruvate, which can undergo oxidative decarboxylation yielding acetyl–S.CoA, while others give acetyl–S.CoA directly. In the end, therefore, fats, carbohydrates and proteins alike are degraded in one way or another, depending on the group of substances concerned, *but all arriving in the end at the formation of acetyl–S.CoA*. The later adventures of this remarkable substance we shall consider in the next chapter. In the meantime we can conveniently summarize the main metabolic lines we have reviewed in the form of the diagram shown in Fig. 14.

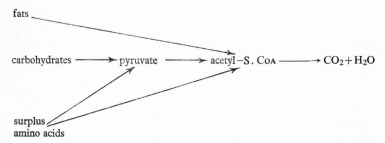

Fig. 14. Summary of main pathways of metabolism of the principal foodstuffs and their derivatives.

THE POWER-HOUSE OF THE CELL

THE CITRIC ACID CYCLE

Only a small part of the energy required by a living cell comes from the breakdown of carbohydrate, fat and surplus amino acids to acetyl–S.CoA. Three-quarters or more of the whole comes from the breakdown of acetyl–S.CoA itself and we shall consider next the underlying mechanisms. In this chapter we shall not attempt to describe the historical development of what has come to be known as the *citric acid cycle*, through which the acetyl–S.CoA is finally broken down to CO_2 and water. Owing to its complexity, we shall describe the citric cycle only in outline, and the manner in which it is believed to operate.

This important mechanism operates almost exclusively in the minute intracellular particles known as *mitochondria*, so we must look there for the main sources of cellular power. The first step in the process, which incidentally was the last to be experimentally demonstrated, consists in a reaction between acetyl–S.CoA and oxaloacetic acid, the latter reacting in its enol form:

$$
\begin{array}{cc}
\text{COOH} & \text{COOH} \\
| & | \\
\text{CH}_2 & \text{CH} \\
| \quad \rightleftharpoons & \| \\
\text{CO} & \text{C(OH)} \\
| & | \\
\text{COOH} & \text{COOH} \\
\text{(keto)} & \text{(enol)}
\end{array}
$$

oxaloacetic acid

The first step can be written as follows:

$$
\begin{array}{l}
\text{COOH} \\
| \\
\text{CH} \\
\| \\
\text{C(OH)COOH} \\
\quad + \\
\text{CH}_3 \\
| \\
\text{CO.S.CoA}
\end{array}
\quad \xrightarrow{+\,H_2O}
\begin{array}{l}
\text{COOH} \\
| \\
\text{CH}_2 \\
| \\
\text{C(OH)COOH} \quad + \quad \text{CoA.SH} \\
| \\
\text{CH}_2 \\
| \\
\text{COOH}
\end{array}
$$

acetyl–S.CoA   *citric acid*

The Citric Acid Cycle

Thus the first step in the degradation of acetyl–S.CoA consists in its synthesis, with the aid of oxaloacetate, into the six-carbon compound, *citric acid*. At the same time the coenzyme A is liberated to be used over again in the formation of further molecules of acetyl–S.CoA.

Once formed the 6-carbon citric acid undergoes a series of enzyme-catalysed changes in the course of which a molecule of CO_2 is split off, leaving a 5-carbon compound which we have already encountered as the product of transdeamination of glutamic acid, viz. *α-ketoglutaric acid*. From this a second molecule of CO_2 is removed by oxidative decarboxylation (p. 77) yielding *succinic acid*, and by this stage the two carbon atoms entering the system in the form of the two carbons of the acetyl residue of acetyl–S.CoA have undergone oxidation to CO_2. In outline the reactions are as follows:

$$
\begin{array}{ccc}
 & \begin{array}{c} CO_2 \\ + \end{array} & \\
\text{COOH} & \text{COOH} & \\
| & \cdots|\cdots & \text{COOH} \\
\text{CH}_2 & \text{CO} & | \\
| & | & \text{CH}_2 \\
\text{C(OH)}\!:\!\text{COOH} \longrightarrow & \text{CH}_2 \longrightarrow & | \\
| & | & \text{CH}_2 \quad +CO_2 \\
\text{CH}_2 & \text{CH}_2 & | \\
| & | & \text{COOH} \\
\text{COOH} & \text{COOH} & \\
\text{citric} & \text{α-ketoglutaric} & \text{succinic} \\
\text{acid} & \text{acid} & \text{acid}
\end{array}
$$

These are only much abbreviated expressions of the reactions that take place; the stage from citrate to α-ketoglutarate involves several intermediate steps, all of which are known and can be demonstrated independently under suitable conditions. The further oxidative decarboxylation of α-ketoglutarate is likewise a rather complicated stepwise process and here again the intermediate steps can be independently demonstrated under suitable experimental conditions.

However, this is not the whole story. Oxaloacetate is required to start the whole process, so that a constant supply of oxaloacetate would be required unless, indeed, it could be regenerated and used over again, as so commonly happens in biochemical systems. We have seen that cyclical processes are often used in metabolic operations, as for instance, in the synthesis of urea (p. 53), in the transdeamination of amino acids (p. 50) and again even in photosynthesis (p. 69). Something of the same general kind happens here again.

As we saw, the liberation of the second molecule of CO_2 is attended by the formation of *succinate* and, from this succinate, oxaloacetate is re-formed through a series of comparatively simple reactions:

$$
\begin{array}{ccccccccc}
\text{COOH} & & \text{COOH} & & \text{COOH} & & \text{COOH} & & \text{COOH} \\
| & & | & & | & & | & & | \\
\text{CH}_2 & & \text{CH} & & \text{CH}_2 & & \text{CH}_2 & & \text{CH} \\
| & \xrightarrow{-2\text{H}} & \| & \xrightarrow{+\text{H}_2\text{O}} & | & \xrightarrow{-2\text{H}} & | & \rightleftharpoons & \| \\
\text{CH}_2 & & \text{CH} & & \text{CHOH} & & \text{CO} & & \text{C(OH)} \\
| & & | & & | & & | & & | \\
\text{COOH} & & \text{COOH} & & \text{COOH} & & \text{COOH} & & \text{COOH} \\
\textit{succinic} & & \textit{fumaric} & & \textit{malic} & & \text{(keto-)} & & \text{(enol-)} \\
\textit{acid} & & \textit{acid} & & \textit{acid} & & \multicolumn{3}{c}{\textit{oxaloacetic acid}}
\end{array}
$$

The enzymes catalyzing these reactions are well known and are present in cells of practically every kind, in animals, plants and many kinds of micro-organisms. The first enzyme, *succinate dehydrogenase*, catalyses the transference of 2H to an appropriate H-acceptor, leaving fumarate, which undergoes hydration at the hands of *fumarase*, and yields malate. Malate then undergoes dehydrogenation under the influence of a *malate dehydrogenase*. The product, *oxaloacetate*, is produced in the keto-form and enolizes, and this last step is probably the only reaction in the whole system for which no enzyme is required, being, as it is, a simple and probably spontaneous intramolecular rearrangement.

The system as a whole can now be summarized in the diagrammatic form shown in Fig. 15.

In effect, as this diagram makes clear, a molecule of acetate, in the form of its highly reactive CoA.SH derivative, is oxidized at each turn of the cycle, two molecules of CO_2 are formed, and the oxaloacetate molecule required to introduce the next acetyl group into the cycle is regenerated. Theoretically, therefore, a single molecule of oxaloacetate can be used over and over again to bring about the oxidation of a theoretically unlimited number of acetate residues.

The over-all equation for the operation of the cycle is as follows, if we ignore the CoA, which acts only as a catalyst and is regenerated at the end of the reaction;

$$CH_3COOH + 2O_2 \rightarrow 2CO_2 + 2H_2O.$$

So far we have accounted for the production of the two molecules of carbon dioxide, and the next problem to be considered is the origin of the oxygen.

THE RESPIRATORY MECHANISMS

The reader will probably have noticed that the instances of biological oxidation mentioned in these pages have always been described as involving dehydrogenation of the substrate and the transference of

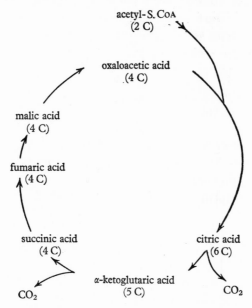

Fig. 15. Outline of the citric acid cycle. See text for further details.

2H to some appropriate acceptor. The general reaction for such processes can be written thus:

Here AH_2 is the substance undergoing oxidation and B is the hydrogen acceptor. This, in fact, is the usual machinery for biological oxidation; very few cases are known in which oxygen is used directly as an oxidizing agent. However, if pairs of hydrogen atoms were continuously parked on the natural hydrogen acceptors of the cell, a time would come when no parking space remained and cellular oxidations would come to an untimely end—untimely for the cell.

In actual practice, however, the situation is dealt with by having a series of hydrogen acceptors arranged in a chain-like manner, each taking over H atoms from the one immediately preceding it in the chain. Each acceptor in turn is reoxidized by passing on its burden of H atoms to the next member of the chain. Eventually all the hydrogen atoms, whatever their origin, are passed on to an intracellular pigment known as *cytochrome*, which differs from all the other hydrogen acceptors in that its reduced form can be reoxidized by molecular oxygen under the influence of an almost universally distributed enzyme known as *cytochrome oxidase*. Here, at the very end of the chain, *oxygen* at last comes into the picture, and this is the fate of oxygen carried to the cells by way of the blood. Indeed, the main function of the oxygen we breathe is not that of oxidizing the food we eat but of *keeping our cytochrome in the oxidized condition*.

Diagrammatically the respiration chains can be represented as follows, where AH_2 represents the original hydrogen donor, X and Y two of the intermediate carriers which conduct hydrogen from the substrates of oxidation at one end of the chain to cytochrome at the far end:

$$AH_2 \quad X \quad YH_2 \quad \text{oxidized} \quad H_2O$$
$$\text{cytochrome}$$
$$A \quad XH_2 \quad Y \quad \text{reduced} \quad \tfrac{1}{2}O_2$$

dehydrogenase   *cytochrome oxidase*

(The dotted lines are put in to indicate that additional carriers other than X and Y play a part in the systems as they occur in nature.)

Here, as in so many other biochemical phenomena, we have instances in which a single molecule of some substance, such as X, Y and cytochrome itself can be used over and over again, bringing about the chemical transformation of a theoretically unlimited number of molecules of the substrate the system is designed to attack.

Let us now see how these respiratory chains co-operate with the enzymes of the citric acid cycle. There are, in the complete citric system, four steps at which dehydrogenation takes place. The first lies on the route *between citrate and α-ketoglutarate,* the second in the course of the *oxidative decarboxylation of α-ketoglutarate,* the third in the *oxidation of succinate to fumarate,* and the fourth in the *oxidation of malate* back to oxaloacetate. In all, therefore, four pairs of hydrogen atoms are fed into the carrier chains. They pass then along the respiratory chains to cytochrome, four molecules of which are reduced and require *four atoms of oxygen* for their re-oxidation. *Four molecules of water* are produced in this last reaction *and, of these, two are consumed* in the reactions of the cycle; one is used in the conversion of fumarate to malate (p. 88) and a second in the oxidative decarboxylation of α-ketoglutarate (p. 87). All in all, therefore, one molecule of acetate is oxidized and four oxygen atoms are used in the process, two molecules of carbon dioxide are formed, together with two molecules of water;

$$CH_3COOH + 2O_2 \longrightarrow 2CO_2 + 2H_2O.$$

And this is the overall equation for the complete oxidation of acetic acid. The process is admittedly a complicated one, but everything adds up.

Recollecting as we must that carbohydrates, fats and superfluous amino acids all undergo metabolic conversion into acetyl–S.CoA we can now realize the importance of the citric acid cycle as a clearing house for the final oxidation of all three of the principal articles of food which make up an average diet. It may be asked why it is that the breakdown of, say, glucose needs to be carried out in the way it is, passing through a long series of changes before it becomes acetyl–S.CoA, and even then has still to be put through the machinery of the citric cycle. This is indeed a most important question and one which we shall presently attempt to answer.

THE UNIT OF BIOLOGICAL ENERGY

We have already raised the question of why the metabolic breakdown of food materials should be as long and tortuous a process as indeed it is. It might be thought that the same result could be obtained simply by throwing food on to a kind of metabolic furnace. The end-products would still be CO_2 and water and the output of energy would be the same, but in this case the living organism would have to function as some kind of heat engine, and heat engines cannot do useful work of any kind except by conducting heat from a higher to a lower temperature. But the animal organism is at much the same temperature in all its parts, and so would be the most grossly inefficient of heat engines if, indeed, it would work at all.

The living cell uses quite different devices. By breaking up the processes into many different serial steps the energy available is not set free all at once as it would be in a furnace, but in small packets of energy at many of the intermediate steps. These parcels can then be gathered up and stored for use as and when they are required.

The trapping agent is *adenosine diphosphate*, ADP, which we have already encountered, and which arises by the breakdown of the triphosphate ATP. To understand the importance of these compounds in the energy transactions of the body, it is necessary to realize that practically every kind of energy expenditure on the part of the body is associated with the breakdown of ATP, leaving ADP behind. When a muscle contracts and does work, the energy for the contraction and the work it does comes from the breakdown of ATP. The discharge of the electric organs of certain fishes also comes from breakdown of ATP and so too does the flash of the firefly. Indeed, dried firefly tails are commercially available for the detection and measurement of very small amounts of ATP, for the light emitted can be readily and very accurately measured. But quite apart from dramatic cases such as these, ATP provides energy for the multitude of synthetic reactions going on constantly in the body. In the synthesis of polysaccharides such as starch and glycogen and in that of fatty acids and fats, energy is required since all these are 'uphill' reactions and it will be remembered that ATP is generated in the earliest stages of photosynthesis (p. 66). Even the synthesis of comparatively small molecules such as hippuric acid and urea call for a good deal of energy. In all these cases synthetases are involved and the energy required comes from ATP. Often ATP is the immediate energy source and, although in some cases an intermediate energy carrier is required, the energy comes from ATP in the end (cf. p. 73).

The ATP molecule is associated with a large amount of available energy. When, for example, its terminal phosphate group is transferred to some second substance—a common device as the first stage in biosynthetic processes—the phosphate carries with it something of the order of 8000 calories per g molecule. It follows from all this that whatever kind of activity, be it mechanical, electrical or chemosynthetic, ATP is consumed and ADP is left behind, and it appears today that the whole business of the metabolic breakdown of food materials is so designed and directed as to produce the richest possible harvest of ATP to replace that used up in energy-consuming operations.

In the breakdown of glucose to pyruvate, one molecule of ATP can be formed for every molecule of pyruvate produced and a further molecule can be formed in the conversion of pyruvate to acetyl–S.CoA. In the citric acid cycle however there is only one step at which an ATP molecule can be formed. If we consider these reactions alone the outlook for the animal looks very bleak indeed, but the situation is not as bad as appears at first sight. The bulk of the ATP formation going on is associated not so much with the direct lines of metabolic breakdown as with the systems that carry away H atoms from the oxidative reactions. At almost every stage where H atoms are transferred from one carrier to the next in the respiratory chains a new ATP molecule can be formed.

CAPTURE AND STORAGE OF ENERGY

It is possible to measure experimentally the yields of ATP associated with these H-transfers and to arrive from the results at figures for the efficiency of energy-capture, expressed in terms of the number of ATP molecules formed. In the oxidative breakdown of glucose for example, one molecule of ATP can be formed by a reaction on the direct chain, but for each pyruvate molecule produced an oxidative reaction has to take place. This sends off two H atoms along a chain of respiratory carriers and produces three additional molecules of ATP in the process. The subsequent oxidative decarboxylation of pyruvate to give acetyl–S.CoA yields in all four more ATP molecules—total eight.

Going on now into the citric acid cycle there are four reactions involving dehydrogenation and an energy yield corresponding to twelve more molecules of ATP so that, starting with one molecule of glucose (each molecule gives two of pyruvate) and carrying its breakdown right through to CO_2 and water, the total overall yield of ATP amounts to no less than forty molecules or thereabouts for each molecule of glucose oxidized.

Nor does the matter end here. Muscle tissue and the electric organs of fishes are characterized by the almost explosive nature of their responses to stimulation, and during the initial phase of their activity they expend energy faster than it can be produced by the metabolic machinery, for the latter takes time to get under way. Having by now learned something of the ingenuity of living cells, the reader will not be surprised that cells such as those of muscle and electric tissue make provision for this lag period by actually storing up energy produced during rest and relaxation. Although the storage material differs in different groups of animals, the mechanism involved is always much the same. In the case of mammalian and, indeed, all vertebrate muscles and electric organs there is present a simple substance known as phosphocreatine ('phosphagen'):

$$
\underset{\text{phosphocreatine}}{\overset{\displaystyle\text{NH}.\text{PO}_3\text{H}_2}{\underset{\displaystyle\underset{\displaystyle\text{CH}_3}{|}}{\overset{\diagup}{\underset{\diagdown}{\text{HN}=\!\!\!=\text{C}}}}}\;\;+\;\;\text{ADP}\;\;\rightleftharpoons\;\;\underset{\text{creatine}}{\overset{\displaystyle\text{NH}_2}{\underset{\displaystyle\underset{\displaystyle\text{CH}_3}{|}}{\overset{\diagup}{\underset{\diagdown}{\text{HN}=\!\!\!=\text{C}}}}}}\;\;+\;\;\text{ATP}
$$

NH.PO$_3$H$_2$ / HN=C \ N.CH$_2$COOH | CH$_3$ — *phosphocreatine*

NH$_2$ / HN=C \ N.CH$_2$COOH | CH$_3$ — *creatine*

This compound reacts readily and reversibly, under the influence of an appropriate enzyme, with ADP giving creatine and ATP.

In the earliest stages of activity, ATP is broken down to supply the energy being expended but the residual ADP has no need to wait for the whole metabolic machine to awaken; instead it can at once be converted back into ATP at the expense of the energy stored in the form of phosphagen, leaving free creatine. The supplies of storage material seem to be nicely calculated to tide the muscle over until its metabolic mills have got well under way and can produce energy and ATP as fast as they are expended. Finally, during rest, ATP produced by the usual metabolic machinery, reacts with the residual free creatine and builds up phosphagen again, while the residual ADP is fed back to other metabolic systems for rephosphorylation.

It seems remarkable indeed that all these events and many more can proceed at the rate and with the efficiency they do under such moderate conditions of pH and temperature. By the employment of its ordered and organized array of enzymes the living cell can accomplish feats which surpass even those of the chemist with all the artifices of the laboratory at his disposal.

12

NUCLEIC ACIDS AND
NUCLEOPROTEINS

INTRODUCTION

Nucleoproteins are found in cells of every kind and consist of a complex substance called nucleic acid in combination with a basic protein. Relatively little interest has been taken in the protein components, but the nucleic acids have become the centre of a vast, new and exciting field of research. They are perhaps the most important macromolecules found in living systems and were first isolated from tissues rich in nuclei such as the soft roes (milt) of fishes, some 50–80% of the dry weight of which consists of nucleoprotein material. They received their name because of their particular association with cell nuclei, though they are present outside the nucleus too.

The early preparations of these substances were crude, rather insoluble and generally unattractive from the point of view of the pure chemist, so that detailed information about their structure and function was slow in coming. The present phase of interest only began with the rather startling discovery that *certain plant viruses can be isolated in crystalline form* and are, in fact, nucleoproteins. Many of the viruses that attack plants, animals and micro-organisms have now been found to be crystallizable nucleoproteins.

Virus diseases can be transmitted from a sick to a healthy plant by innoculation with the infected sap. The amount of virus present increases as the disease runs its course and may reach 10% or more of the dry weight of the tissues by the time the ailing plant dies. Because viruses can not only transmit disease from sick to healthy organisms but also can multiply, they were formerly regarded as analogous to bacteria, except in size, but unlike most bacteria they can only multiply in the tissues of their own particular hosts and appear to have no independent metabolism of their own. They are so small that they can pass freely through filters that retain ordinary bacteria and were usually referred to as 'filter-passing viruses' and considered to be micro-micro-organisms. The discovery that they can be crystallized came therefore

95

as something of a shock, and not only to the bacteriologists and clinicians.

Virus diseases can be transmitted by the crystalline virus material as well as by infected sap, and again the amount of recoverable virus increases as the disease develops. Here, therefore, we have something that has some of the properties of living material; like many bacteria, the viruses can transmit disease from one organism to another and, again like bacteria, they can multiply in the tissues of their hosts. Yet, at the same time and unlike bacteria, they seem to have no metabolism of their own and, unlike any other kind of living stuff whatsoever, they are crystallizable.

These substances thus bridge the gap between the living and the non-living worlds, a discovery that has had profound effects upon biological thought. It has been necessary, for example, to revise our ideas about the nature and the origin of life; indeed it seems that we can no longer use the word 'life' as a precise term because we now know less than ever exactly what we mean by it.

When presently it was discovered that *chromosomes* too are made up of nucleoprotein material, excitement became intense. Like the viruses, they reduplicate, though in a more controlled manner; moreover, the chromosomes pass on not *disease* from one organism to another, but *hereditary characteristics* from parents to offspring and do so with astonishing accuracy from cell to cell and from generation to generation. Viruses also reproduce themselves with remarkable accuracy. What, then, does the biochemist have to say about these fascinating substances and their equally fascinating performances?

COMPOSITION OF NUCLEIC ACIDS

Just as proteins are built up by stringing together large numbers of amino acids and differ one from another according to the choice of amino acids and the sequence in which they are assembled, nucleic acids are built up from building units called *nucleotides*. The chief difference lies in the number of building blocks available; about twenty amino acids are concerned in the production of proteins, but only four nucleotides are at all common in the nucleic acids. Nevertheless, an immense variety of different nucleic acids can be formed by varying the number and sequence of several hundreds or thousands of the few nucleotides that are available.

Composition of Nucleic Acids

Each individual nucleotide consists of a phosphate residue, a pentose sugar and a nitrogenous base. The base is linked to the glycosidic (reducing) group of the sugar, and the phosphate group to carbon 5' of the sugar ring:

base–pentose–phosphate.

Individual nucleotides can be joined together in larger or smaller numbers, forming chains known collectively as *polynucleotides*. The inter-unit link is formed by a secondary attachment of the phosphate group of one nucleotide to carbon 3' of the pentose residue of the next in the chain:

$$\begin{array}{c} | \\ \text{base–pentose–phosphate} \\ | \\ \text{base–pentose–phosphate} \\ | \\ \text{base–pentose–phosphate} \\ | \end{array}$$

Total hydrolysis of nucleic acids from various sources has made it clear that two classes of these substances exist and that the principal difference between them lies in a difference between the pentose sugars they contain, although there are also minor differences between the bases. In one group the pentose is D-*ribose* and in the other its place is taken by D-2-*deoxyribose* (formulae on p. 64). The two groups are accordingly known as *ribonucleic* and *deoxyribonucleic acids* respectively, usually abbreviated as RNA and DNA.

Practically all of the DNA of most cells is concentrated in the nucleus; RNA, on the other hand, is distributed throughout the cell. Some RNA is present in the nucleus but it is chiefly associated with minute intracellular particles known as *microsomes*, some of which are named *ribosomes* because they are especially rich in RNA. These are much smaller particles than the mitochondria mentioned previously (p. 86); indeed they are so small as to be submicroscopic, though they can be collected and studied by appropriate methods.

This, perhaps, is a good point at which to mention a technique known as *differential centrifugation*, which has played a part of the greatest importance in biochemical research in the last dozen years. Cells or tissues are first ground in a special mill called an homogenizer until the cell walls have been disrupted and the nuclei and other intracellular bodies have been set free. The homogenized material is then centrifuged at a moderate speed for a short time, the conditions being chosen so as to cause the tissue debris to sediment out. Usually the *cell nuclei* come down in the same fraction but can

97

be separated from debris for separate study. After dilution the debris-free material is centrifuged longer and faster to bring down the *mitochondria* and, after these have been removed, the centrifuge is run faster and longer still to bring down the *microsomes*, leaving behind only the now particle-free *cytoplasm*.* With the help of this method it has been possible to learn a great deal about the intracellular localization of many enzymes and other substances, including DNA and RNA themselves.

By hydrolysis with the help of suitable enzymes the two groups of nucleic acids can be dismantled to give their component nucleotides and these in their turn to give the pentose, phosphate and the nitrogenous bases. Some mention has already been made of the sugars. As for the bases, these all have flat or plate-like molecules and fall into two groups, one of which contains the pyrimidine and the other the purine ring:

pyrimidine ring   purine ring

The products obtained by total hydrolysis of DNA and RNA are listed in Table 7.

Table 7. *Products of hydrolysis of ribonucleic and deoxyribonucleic acids*

	RNA	DNA	No. of molecules
Pentose	Ribose	Deoxyribose	4
Acid	Phosphoric	Phosphoric	4
Pyrimidines	Cytosine	Cytosine	1
	Uracil	Thymine	1
Purines	Adenine	Adenine	1
	Guanine	Guanine	1

* Starting with a 10% homogenate in isotonic saline or sucrose and working at 0–2 °C so as to prevent denaturation of the proteins, inactivation of enzymes and the like, 10 min centrifugation in a gravitational field about 500 times that of gravity (500 g) suffices to bring down tissue debris and nuclei and a further 10 min at 8000 g brings down the mitochondria. Another 30 min at 20,000–50,000 g separates the microsomes. A high-speed refrigerated centrifuge is necessary. The mitochondria are osmotically fragile bodies and in order to get them in an unbroken state it is necessary to carry out the fractionation in a solution of sufficient osmotic strength. Sucrose or saline can be used.

Composition of Nucleic Acids

According to the results shown in the Table, the simplest possible formula for a nucleic acid would be that of a *tetranucleotide* and, in fact, this was the accepted formula for a good many years. The analytical figures for the base composition of nucleic acids from many sources do not support this however, and we now know that the molecular chains of the nucleic acids are very long, thin, filamentous or thread-like objects containing very large numbers of nucleotide units, all joined together in the following manner:

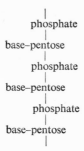

The –pentose–phosphate– arrangement is many times repeated to form a backbone-like main chain, leaving all the bases sticking out on one side. While this is the simplest kind of formula we can give for either RNA or DNA it is by no means the end of the story. There are other significant differences between the two types of nucleic acids, but what follows concerns DNA almost exclusively.

STRUCTURE OF DNA

X-ray analysis is a method beloved of those who study macromolecules and its contributions to our knowledge of the precise, three-dimensional configuration of the DNA molecule has been one of its most brilliant triumphs, but we cannot go into details here. Most brilliant of all was the work of Watson and Crick. In Astbury's hands this method has also given us detailed information about the structures of many simpler natural fibres such as silk, hair, cotton and so on, and even the complex molecule of haemoglobin has given up its structural secrets to the hands of the X-radiographers.

In the case of DNA we now know, among other things, the molecular diameter and the distances along the fibre at which particular features recur. Information like this enabled Watson and Crick to construct scale models of the molecule which could be back-checked against X-ray measurements made by Wilkins. Several such models have been made, but only one satisfies all the requirements specified by the X-ray data.

The principal features of the DNA molecules are, first, that the –pentose–phosphate– chain does not lie in a straight line but in the form of a helix,* i.e. it is coiled in a spiral fashion like an open spring or a spiral staircase. Moreover, there are two such chains forming a double or two-stranded spiral. This arrangement is illustrated by the diagram in Fig. 16.

The bases form a core that lies within the two spirals and are held together in pairs by hydrogen bonds; there are ten of these hydrogen-bonded pairs for each complete turn of the spiral. Although individually the hydrogen bonds are very weak, the collective force exerted when large numbers are present suffices to hold the two helices and their bases together and to give cohesion to the whole structure.

The X-ray data also show that the diameter of the molecule is very constant throughout the fibre, and this condition can only be fulfilled if each pyrimidine base lies opposite to a purine and vice versa. Watson and Crick have been able further to show that adenine (A) always lies opposite to thymine (T) and that guanine (G) and cytosine (C) similarly always lie opposite each other. It is noteworthy in this connexion that DNA normally contains equivalent quantities of adenine and thymine.

Fig. 16. Diagram to show the double helical strands which form the outer portion of the DNA molecule. The core is occupied by the purine and pyrimidine bases.

If now we represent the deoxyribose and phosphate radicals by D and P respectively, and the bases by their initial letters, we can represent the structure of the DNA molecule using dotted lines to represent the hydrogen bonds, as shown at the top of the following page. This general formula carries several implications. First, it represents only a small part of a complete molecule of DNA. Secondly, as we read upwards from the bottom to the top of this diagram, the ascent

* Plural = helices.

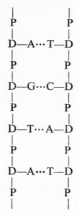

is like that of a spiral staircase rather than that of a ladder: the diagrammatic representation in Fig. 17 will help to make this point clear. Lastly, it should be emphasized that the base pairs are

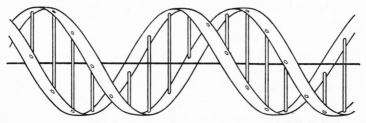

Fig. 17. Diagram to illustrate spatial configuration of the DNA molecule. The intertwining ribbons represent the pentose–phosphate chains forming the two helices of the DNA molecule. The 'rungs' of the ladder each represent a pair of hydrogen-bonded bases which fasten the helices together and form the core of the molecule. (Modified after Watson and Crick.) *To be viewed sideways.*

arranged in many different sequences in DNA obtained from different sources; the pairs used in the type formula above and the sequence in which they are arranged are purely arbitrary.

REPLICATION (REDUPLICATION) OF DNA

Further dramatic developments came with discovery that chromosomes consist mainly of DNA. Before cell division takes place it is usual for the chromosomes to divide into pairs, one of each pair passing into each daughter cell when the parent cell divides. This production of new chromosomal material is accomplished with remarkable precision and accuracy; each new chromosome is a faithful and exact replica of the old and from the first to the last cell division every new chromosome produced has exactly the same composition as the old. But it seems unlikely that the complex molecules of DNA can produce new ones by any simple, straightforward process of 'copying'.

However, enzymes have been discovered in bacterial and animal cells which can catalyse the synthesis of polynucleotides resembling DNA from deoxynucleotide diphosphates. This cell-free system requires small quantities of preformed polynucleotide as a 'starter'. Synthetic polynucleotides have been formed with the aid of these enzymes, some containing all four of the nucleotides found in DNA and others containing nucleotides in which only one particular base is present, e.g. adenine or uracil. These one-base nucleotides are not naturally occurring substances but they have considerable interest and are commonly referred to as 'poly-A' and 'poly-U'.

These synthetic polymers contain the usual –pentose–phosphate– chains but they are not organized into the double-spiral formation found in DNA; they have instead only a random configuration. If, however, poly-A and poly-U are mixed together, their –pentose– phosphate– chains wind around each other in spiral fashion and produce something resembling DNA; a double spiral with the bases forming a core within.

These experiments show two things clearly. First, nucleotides can be made to undergo polymerization and, secondly, polynucleotide chains tend, apparently spontaneously, to intertwine one with another to form doubly stranded helical structures similar to those present in DNA. Given these facts we might expect that if poly-A were prepared and poly-U formation was initiated in its presence by adding the appropriate enzymes and starting materials, we should have a system in which doubly stranded spiral molecules would be built up step by step as each new nucleotide molecule was incorporated. The newly added polynucleotide chain contains uracil and would not, of course, be a copy of the adenine-containing chain from which we started, but all this suggests a way in which DNA might reduplicate without direct 'copying' when the chromosomes divide.

How the formation and replication of new doubly stranded helical molecules like DNA can take place presents many problems, but for purposes of argument we will assume first of all that the two helices present in the parent DNA are in some way unwound or

Replication (*Reduplication*) *of DNA*

untwisted. The two separated helices would not of course be identical, nor would they be mirror images each of the other, but they would be *complementary* in as much as they and only they or exactly similar molecules could fit together and form DNA again. If we call the two separated helices A and B, then, in the presence of the appropriate enzymes and the raw materials for fresh nucleotide synthesis, it seems very probable that A might induce the formation of a new complement* for itself and that B would do likewise. The result would be the formation of two new molecules of AB, each identical in composition with the DNA from which we began.

Leaving aside the question of mechanisms for the moment it is worth while to notice that the replication of DNA does indeed appear to involve something of the kind suggested in the last paragraph. If a suitable micro-organism is grown in the presence of a suitable isotopic label it will synthesize DNA with label in both its helices, which we can represent as **AB**. If the scheme we have outlined has any reality the following pathways would be followed in subsequent generations if the organism is then transferred to an unlabelled medium:

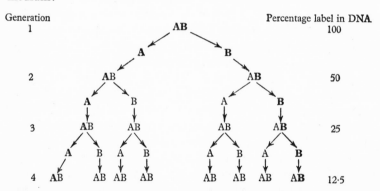

Generation 1 — AB — Percentage label in DNA 100

2 — AB — AB — 50

3 — AB — AB — AB — AB — 25

4 — AB — AB AB AB — AB AB AB **AB** — 12·5

Experiments have been carried out with bacteria grown in a medium containing heavy nitrogen, which was incorporated into their DNA. When grown on for several generations on unlabelled media the DNA at each generation was found to contain the heavy label in just the amounts that would be expected from this scheme. This certainly suggests that each of the two parent spirals does indeed act as a foundation for the formation of one that is complementary to it.

* The term 'complement' is used here in a sense different from that in immunochemistry, and the two must not be confused.

Next let us consider how complement formation might take place. The chances are that if the two helices undergo separation they would lose their helical configuration like the poly-A and poly-U already mentioned. Adenine units in A would tend to become associated by hydrogen bonding with nucleotides containing thymine, since we know that these two lie opposite one another in the DNA molecule. Thymine units in A would similarly tend to associate with nucleotides containing adenine and so on. As adjacent sites on A became occupied, polymerization of the newly added nucleotides might begin and localized patches of double helical configuration would appear. Alternatively, addition of new nucleotides might start at one end of A and spread to the other. In either case this process could continue until eventually all the available sites on A were occupied, polymerization of the added nucleotides would be completed and a complete new double helix would be formed.

The foregoing argument assumes the availability in the cell of supplies of all the nucleotides required to form complement, but this is an altogether reasonable assumption. And, of course, the processes leading to the formation of complement for A would serve equally for the production of complement for B.

Specificity of a high order would certainly be needed in the siting of the newly added nucleotide units if the final product is to be a replica of the DNA molecule from which we started, but specificity of this kind is perhaps not more remarkable than the specificity we find among enzymes.

All that has been said so far depends upon the assumption that the DNA molecule is in some way untwisted to allow separation of two helical strands, but there is no compelling reason to believe that such an untwisting does indeed take place, nor do we know how, if it does take place, the untwisting is accomplished. Much that has been said in the last few pages is therefore purely speculative.*

We have a certain amount of experimental evidence. Most of the work on the replication of DNA in living cells has been done *in vivo* with micro-organisms of various kinds and with viruses that attack plants, animals and bacteria. From this we can at present only guess

* The interested reader will find a wealth of further information in *New Biology*, **31**, a Penguin publication. A number of excellent articles on these and related fields have appeared in *Scientific American* from time to time, any or all of which provide profitable reading.

at the method of replication of *chromosomal* DNA and hope to guess right. Most of the bacterial viruses have double-stranded DNA but some have only a single strand in the molecule; some of the animal viruses contain both DNA and RNA, and plant viruses usually have RNA only.

It is now a truism to say that DNA is the chemical basis of heredity, except for the plant viruses (which do not contain it). But the science of genetics was well established long before anything was known about the chemical nature of the chromosomes and the DNA of which they are made. It is generally believed today that the genes, each of which contributes some hereditary characteristic to a given individual, are strung out along the chromosomes, suggesting that each gene corresponds to a longer or shorter fragment of the DNA molecule with some particular and specific sequence of nucleotide units. Red hair, black skin, blue eyes and the like are all familiar examples of the end-products of inheritable characteristics transmitted by the genes.

Different individuals are all different, except in cases where the fertilized ovum divides into two, each half following its own separate developmental path so that identical twins are produced. Identical twins are identical because they possess exactly the same genes; exactly the same chromosomes. But apart from this, different individuals possess different proteins, different enzymic equipment and so on; it is the cumulative effect of these genetic differences, plus his experience after birth, that makes up the individuality of the individual.

It is usual in the experimental study of genetics to work on very small organisms because they get through a very large number of generations in a short space of time. The fruit fly, *Drosophila*, has been a favourite experimental tool for geneticists for many a year, but more recently even smaller and still more rapidly reproducing organisms, notably the bread mould, *Neurospora*, and a considerable variety of bacteria and viruses, have been used extensively in the study of inheritance and the phenomenon of mutation.

In the ordinary way chromosomes reproduce themselves with 100 % accuracy. Now and again, however, usually as a result of exposure to some kind of radiation, the accuracy falls below the 100 % level and the offspring have features which their parents did not possess, and these new features may be transmitted to the next generation. The majority of these alterations or mutations, are lethal, and so are not handed on, but there are some that are advantageous and these are the basic ingredients for evolutionary advancement. Evolution, in fact, seems to have proceeded by the accumulation of advantageous mutations and, if we wish to study evolution from the biochemical point of view, it is well to study mutation from the same standpoint.

Mutation takes place spontaneously every now and again, but the number of mutations produced in a population can be greatly increased by exposure to X-rays and other forms of radiation. One of the greatest dangers we all face today is the production of mutations through exposure to radioactive fall-out from nuclear explosions. But let us return to more wholesome matters.

Experimental work on *Neurospora* and other micro-organisms has shown that the genetic changes we call mutations result in changes in the enzymic equipment of the progeny. Nearly always the change is an unfavourable one and, broadly speaking, if one gene is damaged or destroyed, the result is the loss, or at least some unfavourable modification, of at least one enzyme. Often the enzyme is one that is ordinarily employed in the production of some substance of metabolic importance, and in such a case the organism can no longer survive unless it is provided with the product it can no longer make for itself.* In the case of mutant variants of *Neurospora* or bacteria this can usually be done rather easily because the organism can be grown on media supplemented with the missing metabolic intermediate, and a vast mass of information has been gathered together about mutation in this way.

Genetic abnormalities are known too in the human species and some are handed on from generation to generation. Albinism, due to the lack of an enzyme that produces dark pigments from the amino acid tyrosine, is one such abnormality. Another is haemophilia, due to the absence of one of several enzymes concerned in the clotting of blood. There are many others.

In Fig. 18 are shown some of the metabolic pathways followed by the amino acid phenylalanine, a group of reactions catalysed by enzymes that seem to be curiously prone to genetic loss or modification. A number of these hereditary disorders are known,

* This may well have been the first step on the way towards symbiosis and parasitism.

each corresponding to blocking of one or other of the normal pathways through lack of the enzyme particularly concerned with a specific reaction. The very rough diagram which follows has been simplified as far as is consistent with intelligibility. Individuals or families are known with metabolic blockages at one or other of the stages A to D on the diagram and the characteristic deficiencies are inherited in accordance with the usual laws of genetics.

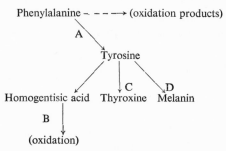

Fig. 18. Metabolic blocks in the metabolism of phenylalanine and tyrosine. Blockage at A, B, C and D lead to imbecilitas phenylpyruvica, alkaptonuria, goitrous cretinism and albinism, respectively.

Phenylalanine is normally converted into tyrosine, but if this pathway is blocked (A) it gives rise by oxidation to a substance called phenylpyruvic acid. This process (A) is blocked in individuals suffering from a certain form of insanity with which is associated the presence of phenylpyruvic acid in the urine. This condition is known as *imbecilitas phenylpyruvica* or *phenylketonuria*. Another block can occur at the tyrosine stage though this block may be only a partial one. It is reflected by the presence of tyrosine in the urine (tyrosinosis) but only one case of this has been reported. Tyrosine itself is normally metabolized through several different channels, one of which goes through an intermediate called homogentisic acid and leads normally to CO_2 and water in the end. In a curious condition known as *alkaptonuria*, the enzyme responsible for breaking down the homogentisic acid is lacking so that homogentisic acid appears in the urine (B). The impressive feature of this disorder is that the urine goes inky black on standing. This is due to the oxidation of the homogentisic acid to a dark coloured quinone, a reaction that takes place especially rapidly under alkaline conditions. Favourable conditions are provided when the urine is allowed to stand so that bacteria get in and break down urea into ammonia.

Two more of these curious metabolic blocks are worth mentioning. Of the several pathways through which tyrosine is normally metabolized, one leads to the thyroid hormone, thyroxine. This path (C) is blocked in some individuals, thyroxin is not formed and the condition called *goitrous cretinism* is the result. Lastly, and through yet other pathways, tyrosine gives rise to melanins, the dark coloured pigments usually present in the skin, hair and certain parts of the eye. This pathway (D) is blocked in *albinism*. All these are the results of genetic abnormalities, mutations which are disadvantageous enough though not actually lethal. All this is compatible with the idea that the chromosomes, the genetic substances, must be in some way concerned with enzyme production and that enzyme production is in some way *controlled* by the chromosomal DNA. What enzymes, proteins and other specialized and specific substances an organism possesses is, apparently, determined by the kind of DNA it has in its chromosomes, but it is doubtful whether the nuclear DNA could ever be directly responsible, alone and unaided, for the formation of all the multitudinous proteins, enzymes and so on that the cell must have if it is to be a going concern.

There is now a general belief that proteins—including of course enzymes—are produced not by DNA itself but by RNA contained in the ribosomes. It is in fact well established that this ribosomal RNA is the most active site of protein synthesis in the cell. Presumably, then, RNA must possess some code or system that enables it to assemble the right amounts of the right amino acids in the right sequence to produce the right, specific kind of protein. There must be many such systems, each with its own code, to account for the production of a large number of different, specific proteins, enzymes included. Yet, according to genetics, all the necessary information is contained primarily in the nuclear DNA, and is passed on from one generation to the next.

If the nuclear DNA carries in its structure a master code containing all the information needed for the synthesis of specific proteins, and carries this code through from one generation to the next, it may be that the DNA acts as a mould or pattern for the synthesis of a 'messenger' RNA in the nucleus and that the latter then enters the cytoplasm and transmits the information impressed upon it to the ribosomal RNA which directs and controls protein synthesis. At all events, recent work has shown that cell-free enzymes can be obtained that will synthesize RNA from the triphosphates of the corresponding nucleotides. It is perhaps significant that these enzymes require the presence of small amounts of DNA to start the synthesis of RNA. Current opinion is that a 'messenger' RNA is formed in the nucleus, receiving its information from the nuclear DNA and communicating it to the ribosomes, where it lays down the pattern necessary for the synthesis of specific proteins.

Thus, although we know a great deal about the synthesis of fats, polysaccharides and a wide variety of other complex substances, our knowledge of protein synthesis is still only at its beginning. At the present time laboratories all over the world are engaged in gathering together information on the synthesis of proteins in cell-free systems, in determining the structure of proteins in minute detail and in attempting to trace the relationships that must certainly exist between the genetic make-up of the cell and the structural features of the specific proteins that the cell elaborates. Biochemists have tackled and solved many difficult puzzles in the past and today they are tackling—and in time will solve—the puzzle of specific protein synthesis. The answer is perhaps not just around the corner, but it may not be very long delayed. In fact, if the present rate of progress is maintained we may well know most of the answers before this new edition appears in the book shops.

INDEX

Index

Index

hexokinase, 34, 72, 74
hippuric acid, 82, 92
histidine, 28
homogentisic acid, 107
hormones, 75, 76
hydrogen, 43
  acceptors, 44, 49, 50, 82–84, 93
  bonds, 95, 96, 99, 100, 104
  ions, 16, 17, 28–32
hydrogen peroxide, 42–4
hydrolases, 43, 45
hydrolysis, 39, 42, 70, 80
  in digestion, 45
  of RNA and DNA, 98

ionic balance, 3 ff.
  composition of tissue fluids and sea water, 5
ionization of amino acids, 33–5
  of blood proteins, 17–18, 30
  of proteins, 33–5
  of water, 31
  of weak acids and bases, 29–35
immunization, 21–2
individuality, 101
inhibitors, 38
insulin, 34, 75, 76
intestinal juice, 46, 70, 80
inulin, 57
iodine, 57
iron, 12
isoelectric point (pH), 32
isomerases, 44, 68, 71
isotopes, 81, 103

katabolism, 45
α-keto acids, 48, 77
α-ketoglutarate, 48–9, 52, 87
ketone bodies, 83
kidney, 54, 70

lactate, 43, 44
lactate dehydrogenase, 43, 44
lactose, 70, 74
levans, 57
life, origin of, 1–3, 96
  nature of, 96
ligases, 44
lipase, 42, 43, 45, 80
liver, 34, 80
  glycogen in, 57, 71–73
  transdeamination in, 48–50
  urea synthesis in, 51–3

lizards, 55
luciferase, 34
lyases, 44, 69, 76
lysine, 26–7, 35
lysozyme, 34

Maia, 7–8
malate, 88, 89
  dehydrogenase, 88, 89
maltase, 44, 70
maltose, 44, 65, 70
mammals, 52–56
mannose, 60, 70
metabolism, 45, 47, 85, 91, 92, 93
  of carbohydrates, 71–78, 85, 91
  of fats, 81–85
  of glucose and glycogen, 76
  of proteins, 21–8, 46–56
methionine, 28
micro-organisms, 36, 70, 95
  in DNA synthesis, 103, 104
microsomes, 97
migrations of animals, 6–10
milk, souring of, 36
  lactose in, 65
mitochondria, 86, 97
molecular weight of proteins, 34
Mollusca, 12
monosaccharides, 45, 58 ff.
mutarotation, 60, 62–3
muscle, 34, 43
  energy for, 90–94
mutation, 4, 101–4

Neurospora, 104, 105
nitrogen balance, 24–5
  excretion, 51–6
nucleic acids, 95–109
  composition, 96–99
  replication, 101–105
  structure, 99–101
nucleoproteins, 95–109
nucleotides, 96–107
nucleus, 93, 109
nutrition, proteins in, 23–6

oleic acid, 79
ornithine, 51–6
  cycle, 52–6
osmotic pressure, 6–10
  of body fluids, 6–10
  of proteins, 32

113